Environmental
MANAGEMENT

Problems and Solutions

Environmental
MANAGEMENT

Problems and Solutions

Edited by

R. Ryan Dupont
Utah Water Research Laboratory
Utah State University, Logan, UT

Terry E. Baxter
Northern Arizona University
Flagstaff, AZ

Louis Theodore
Manhattan College, Riverdale, NY

LEWIS PUBLISHERS
Boca Raton Boston London New York Washington, D.C.

Library of Congress Cataloging-in-Publication Data

Environmental management : problems and solutions / edited by R. Ryan
   Dupont, Terry E. Baxter and Louis Theodore.
         p.   cm.
   Includes bibliographical references and index.
   ISBN 1-56670-316-6 (alk. paper)
   1. Environmental management. I. Dupont, R. Ryan. II. Baxter,
Terry E. III. Theodore, Louis.
GE300.E584 1998
363.7—dc21   98-2785
                                                                                        CIP

PREFACE

Recognizing the need to support undergraduate educators in the development of environmental management educational materials, the National Science Foundation funded a College Faculty Workshop on Environmental Management that was conducted at Utah State University, Logan, in July and August 1996. NSF College Faculty Workshops are designed to "involve college faculty members in preparing course materials and in testing the effectiveness of these science curricular innovations for implementation." To qualify, topics must be of sufficiently broad applicability and impact to warrant national implementation for the enhancement of undergraduate science curricula. In awarding College Faculty Enhancement grants, the NSF gives priority to the development of more efficient and effective educational procedures in newly emerging, interdisciplinary, and problem-relevant subject areas. The principal objectives of the Utah State University seminar on environmental management were: (1) to provide a meaningful course which would generate new ideas and innovative educative approaches in the emerging field of environmental management, and (2) to develop an applications-oriented problem workbook which would support undergraduate faculty involvement in the production of course materials in a range of environmental management areas.

An interdisciplinary group of engineering and science faculty from across the United States involved in undergraduate environmental engineering, chemical engineering, or applied science courses was selected to attend the 2-week seminar. These faculty are listed in the Faculty Contributors section. Lecturers who assisted in the presentation of course material are listed in the Workshop Lecturers section. Each faculty member was required to generate approximately 10 meaningful, applications-oriented problems. The 1996-1997 academic year afforded workshop participants an opportunity to classroom test these problems. A 3-day follow-up session held in June 1997 at Manhattan College, Riverdale, New York, was utilized to revise, update, and edit the final workbook. *Environmental Management: Problems and Solutions* is the product of this year-long effort.

This workbook contains more than 200 problems related to a variety of topics of relevance to the environmental management field. These problems are organized into the following categories: Regulations, Health and Hazard Risk Assessment, Risk Communication, Pollution Prevention, Energy Conservation, Air Quality Issues, Water Quality Issues, Solid Waste Management Issues, Industrial Hygiene, ISO 14000, Native American Issues,

Ethics, Environmental Accounting and Liability, and Other Environmental Issues. Contributions to the workbook from a faculty member carries that individual's initials, e.g., the initials [hb] refer to Howard Beim. Each participant's initials are shown by their name in the Faculty Contributors section.

Detailed solutions to each problem are provided in Section III of each chapter. Solutions are identified with a number and a name corresponding to the original problem given in Section II of each chapter. For example, the solution to the problem "11. Pollution Prevention Problem 11" is found in Section III of that chapter under the title "11. Pollution Prevention Solution 11."

Acknowledgments are due to Tammy Petersen and Ann Kaptanis, who attended to the myriad of details required for the smooth and successful operation of the workshop at Utah State University and the follow-up session at Manhattan College, respectively. The editors also wish to acknowledge the National Science Foundation, without whose support this book would not have been possible.

It is the hope of the editors and contributors to this workbook that *Environmental Management: Problems and Solutions* will provide needed support for those faculty developing courses in environmental management, and that it will become a useful resource for the training of engineers and scientists in mastering this critical topic area.

NSF WORKSHOP PROJECT DIRECTORS

R. Ryan Dupont (Director), Professor and Head, Division of Environmental Engineering, Department of Civil and Environmental Engineering, Utah State University, Logan, Utah. [rrd]. rdupo@pub.uwrl.usu.edu.

Terry E. Baxter (Co-Director), Assistant Professor, College of Engineering and Technology, Northern Arizona University, Flagstaff, Arizona. [teb]. terry.baxter@nau.edu

Lou Theodore (Co-Director), Professor, Chemical Engineering, Manhattan College, Bronx, NY. ltheodore@manhattan.edu.

FACULTY CONTRIBUTORS

Bill Auberle Civil and Environmental Engineering, Northern Arizona University, Flagstaff, Arizona [wma]

Howard Beim Department of Mathematics and Science, U.S. Merchant Marine Academy, Kings Point, New York [hb]

Carolyn Daugherty Department of Geography and Public Planning, Northern Arizona University, Flagstaff, Arizona [cmd]

Bill Doucette Department of Civil and Environmental Engineering, Utah State University, Logan, Utah [wjd]

Kumar Ganesan Department of Environmental Engineering, Montana Tech, University of Montana, Butte, Montana [kg]

Pankajam Ganesan Industrial Hygienist, Butte, Montana [pg]

Michael Haradopolis Brevard Community College, Melbourne, Florida [mh]

Leslie Henry Applied Science and Technology Department, Oglala Lakota College, Kyle, South Dakota [lrh]

Gary Hickernell Division of Natural Sciences, Keuka College, Keuka Park, New York [glh]

Dave James Department of Civil and Environmental Engineering, University of Nevada, Las Vegas, Nevada [dj]

Christopher Koroneos Department of Chemical Engineering, Columbia University, New York [cjk]

F. William Kroesser Environmental Engineering and Science, West Virginia Graduate College, South Charleston, West Virginia [fwk]

Reid Lea Department of Civil Engineering, Louisiana State University, Baton Rouge, Louisiana [wrl]

Sean Xiaoge Liu Department of Civil and Environmental Engineering University of California, Berkeley, California [sxl]

Scott Lowe Department of Environmental Engineering, Manhattan College, Riverdale, New York [sl]

Premlata Menon Environmental and Occupational Health Program, School of Public Health, University of Hawaii, Honolulu, Hawaii [pm]

Monica Minton Department of Chemistry, Marymount College, Tarrytown, New York [mm]

Suwanchai Nitisoravut Inter-Institute Ph.D. Program, Civil Engineering, North Carolina State University/University of North Carolina at Charlotte [sn]

Carol Reifschneider Hagner Science Center, Department of Science and Math, Montana State University Northern, Havre, Montana [car]

Joseph Reynolds Department of Chemical Engineering, Manhattan College, Riverdale, New York [jr]

Myron Robinson Department of Chemical Engineering, Long Island University, Flushing, New York [mr]

Darwin Sorensen Civil and Environmental Engineering, Utah State University, Logan, Utah [dls]

Dave Stevens Civil and Environmental Engineering, Utah State University, Logan, Utah [dks]

Robert Tidwell Natural Resources, Crownpoint Institute of Technology, Crownpoint, New Mexico [rt]

Poa-Chiang Yuan Department of Civil Engineering, Jackson State University, Jackson, Mississippi [pcy]

NSF WORKSHOP LECTURERS

Terry E. Baxter Ph.D., PE, Assistant Professor, College of Engineering and Technology, Northern Arizona University, Flagstaff, Arizona

Anthony Buonicore P.E., CEO, Environmental Data Resources, Inc., Southport, Connecticut

R. Ryan Dupont Ph.D., Professor and Head, Division of Environmental Engineering, Department of Civil and Environmental Engineering, Utah State University, Logan, Utah

David E. James Ph.D., Associate Professor, Department of Civil and Environmental Engineering, University of Nevada, Las Vegas, Nevada

John Wilcox Ph.D., Professor and Chair, Religious Studies Department, Director of the Center for Professional Ethics, Manhattan College, Riverdale, New York

TABLE OF CONTENTS

Preface
NSF Workshop Project Directors
Faculty Contributors
NSF Workshop Lecturers

Chapter 1

REGULATIONS

Darwin L. Sorensen

I. INTRODUCTION

A. THE REGULATORY FRAMEWORK

Environmental management is accomplished principally in response to requirements in environmental laws and regulations. In more recent years "surrogate regulators" and the emergence of the International Organization for Standardizations (ISO) have become important in affecting environmental management approaches. However, federal, state, and local regulations remain the greatest overall influence on the practice of environmental science, engineering, and sociology. Both government and private funding for environmental management activities become available in response to requirements to comply with environmental laws. In this way, environmental regulations become the framework for environmental management.

Environmental management has as its objective the protection of human health and well being, and the protection (preservation and conservation) of life forms and their habitats. Environmental statutes at the federal and state levels, as well as ordinances at the local government level, have this same broad objective. Understanding specific environmental management objectives in each law is critical to effective environmental management practice.

The solution to environmental problems typically involves the cooperation of multidisciplinary and interdisciplinary teams. Scientists, engineers, sociologists, and lawyers frequently work together to design and implement processes or procedures to solve or prevent a real or perceived environmental problem. Legal, economic, natural systems, and engineered systems components may need to be integrated to arrive at an appropriate solution. It is easy for scientists and engineers, who often are focused on applications of science and technology to prevent or remedy environmental damage, to forget that the practice of environmental protection and management draws its authority from the concerns and desires of a community of people. These individuals, for the most part, do not have technical backgrounds and look to their elected representatives and the legal system to protect them from environmental harm. Again, laws and regulations provide the framework in which environmental protection and management are accomplished.

1

B. THE REGULATION SYSTEM

Over the past two decades environmental regulation has become a system in which laws, regulations, and guidelines have become interrelated. Requirements and procedures developed under previously existing laws may be referenced to in more recent laws and regulations. The history and development of our regulatory system has led to laws that focus principally on only one environmental medium, i.e., air, water, or land. Some environmental managers feel that more needs to be done to manage all of the media simultaneously. Hopefully, the environmental regulatory system will evolve into a truly integrated, multimedia management framework.

Federal laws are the product of Congress. Regulations written to implement the law are promulgated by the Executive Branch of government, but until judicial decisions are made regarding the interpretations of the regulations, there may be uncertainty about what regulations mean in real situations. Until recently, environmental protection groups were most frequently the plaintiffs in cases brought to court seeking interpretation of the law. Today, industry has become more active in this role. Forum shopping, the process of finding a court that is more likely to be sympathetic to the plaintiffs' point of view, continues to be an important tool in this area of environmental regulation. Many environmental cases have been heard by the Circuit Court of the District of Columbia.

Enforcement approaches for environmental regulations are environmental management oriented in that they seek to remedy environmental harm, not simply a specific infraction of a given regulation. All laws in a legal system may be used in enforcement to prevent damage or threats of damage to the environment or human health and safety. Tax laws (e.g., tax incentives) and business regulatory laws (e.g., product claims, liability disclosure) are examples of laws not directly focused on environmental protection, but that may also be used to encourage compliance and discourage non-compliance with environmental regulations.

Common law also plays an important role in environmental management. Common law is the set of rules and principles relating to the government and security of persons and property. Common law authority is derived from the usages and customs that are recognized and enforced by the courts. In general, no infraction of the law is necessary when establishing a common law suit. Legal precedent often forms the foundation for common law court actions. A common law "civil wrong" (i.e., environmental pollution) that is brought to court is called a tort. Environmental torts may arise because of nuisance, trespass, or negligence.

Laws tend to be general and contain uncertainties relative to the implementation of principles and concepts they contain. Regulations derived from laws may be more specific, but are also frequently too broad to allow clear translation into environmental technology practice. Permits may be used in the environmental regulation industry to bridge this gap and provide specific, technical requirements imposed on a facility by the regulatory agencies for the discharge of pollutants or on other activities carried out by the facility that may impact the environment.

Most major federal environmental laws provide for citizen law suits. This empowers individuals to seek compliance or monetary penalties when these laws are violated and regulatory agencies do not take enforcement action against the violator.

C. ABOUT THE PROBLEMS IN THIS CHAPTER

The problems in this chapter emphasize regulatory response, but also include issues related to surrogate regulators and surrogate regulation. In addition, an overview of major federal environmental laws and an introduction to some more specific environmental regulatory issues (e.g., takings) are presented through the problems contained in this chapter.

II. PROBLEMS

1. *Regulations Problem 1.* (environmental laws, regulations). [sn]. Briefly describe the major laws that address the various environmental-related issues currently of concern in the U.S.

2. *Regulations Problem 2.* (environmental laws, guidelines, regulations). [dls]. How are the environmental laws, regulations, guidelines and case-specific interpretations by the courts becoming a "system" for managing the environment in the U.S.? Provide at least one example of such a "system" and include your source or reference for this example.

3. *Regulations Problem 3.* (environmental laws, Common Law). [dls]. What is the utility of Common Law in environmental management?

4. *Regulations Problem 4.* (environmental laws, environmental management, risk reduction). [dls]. Explain the relationships among federal environmental laws, environmental management and health risk reduction in the U.S.

5. *Regulations Problem 5.* (environmental laws, environmental management). [dls]. Are U.S. environmental regulations a good initial model for environmental management elsewhere in the world? Provide specific explanations and comments.

6. *Regulations Problem 6.* (environmental laws, "takings"). [dls]. Describe the issue of "takings" under environmental laws. Which laws are most involved in this issue?

7. *Regulations Problem 7.* (environmental laws, ARARs, NCP). [dls]. What are ARARs under the National Contingency Plan? Give an example for a Superfund site.

8. *Regulations Problem 8.* (environmental laws, pollution prevention, Pollution Prevention Act). [dls]. Describe similar features and incentives between existing industrial pollution prevention programs (e.g., 3M's 3Ps) and implementation of the Pollution Prevention Act of 1990.

9. *Regulations Problem 9.* (enforcement, surrogate regulations). [hb]. One of the defining characteristics of the 1990s has been the movement of the federal government towards ever tighter and tighter budgets. This movement has persisted through both Republican and Democratic administrations. One consequence of the budget cuts has been to reduce the ability of the U.S. EPA to pursue enforcement of its regulations. The states also have limited resources for enforcement.

9. *Regulations Problem 9.* [hb] (continued)

Some experts, however, have claimed that certain market forces have served to pressure business to continue observing the regulations. These market forces are called "surrogate regulators." List six possible surrogate regulators and briefly explain how each works towards the enforcement of environmental regulations.

10. *Regulations Problem 10.* (OSHA, regulations, U.S. EPA). [mh]. In a short two paragraph essay describe what the acronym OSHA means and why this law and agency were established.

11. *Regulations Problem 11.* (criteria, hazardous waste, properties, RCRA). [dj]. List the general criteria that, under the Hazardous and Solid Waste Amendments (HSWA) of 1984, are used to define the types of wastes that are hazardous.

12. *Regulations Problem 12.* (criteria, hazardous waste, properties, RCRA). [dj]. List and quantify the four chemical properties against which a non-specific waste is evaluated to determine if it is hazardous as defined by the Resource Conservation and Recovery Act (RCRA) of 1976.

13. *Regulations Problem 13.* (CERCLA, legislation, Superfund). [dj].

  a. Explain the words in the acronym CERCLA.
  b. When and why was CERCLA enacted?
  c. What was the purpose of the Superfund provision of CERCLA?

14. *Regulations Problem 14.* (CERCLA, Superfund, prosecution, liability). [mh]. With regard to CERCLA (Superfund), what are the three main powers that the U.S. EPA possesses in order to prosecute alleged offenders of the Superfund laws?

15. *Regulations Problem 15.* (hazardous waste, solid waste, waste management, waste laws, regulations). [wma]. You and your family have chosen to pursue a new lifestyle and purchased a small farm in the Midwest. While excavating the foundation for a new barn, you discover three buried drums. The drums are nearly full of a dark gray substance, but have no labels or other identifying marks to suggest origin or content.

  a. What federal environmental law or laws may apply to this situation?
  b. As a responsible citizen (but with limited finances), what steps should you take to assure proper disposal of this material?

16. *Regulations Problem 16.* (emergency planning and community right-to-know, EPCRA, SARA Title III, toxic release inventory, TRI). [hb]. One fundamental goal of any Environmental Management System is to prevent releases of harmful chemicals into the environment. Yet, in the past, it was difficult to know if this goal was being met across the nation. Partly as a means to remedy this problem, the Emergency Planning and Community Right-to-Know Act (EPCRA) was passed in 1986. It was originally known as Title III of the Superfund Amendments and Reauthorization Act (SARA) of 1986. One section of EPCRA establishes reporting requirements for the release of a number of chemicals. Chemicals subject to the law include Occupational Safety and Health Act (OSHA) chemicals, hazardous substances under the Comprehensive Environmental Response, Compensation and Liability Act (CERCLA), extremely hazardous substances under EPCRA Section 302, and the chemicals listed as toxic under EPCRA Section 313. The reported releases are compiled by the U.S. EPA into a Toxic Release Inventory (TRI) which is published annually and made available to the public in a computerized database.

    a.   Explain who is subject to the release reporting requirements under EPCRA.
    b.   What constitutes a reportable release?

17. *Regulations Problem 17.* (Superfund, budget). [mh]. Today, with so many concerns involving the federal budget, the CERCLA (Superfund) program has come under significant criticism. Since 1980 only approximately 400 of over 2,100 sites have been remediated and closed, while budget expenditures have reached more than $10 billion. Why have so few sites been cleaned?

18. *Regulations Problem 18.* (Clean Air Act, U.S. EPA, state and local agencies). [mh]. What should small businesses know about the 1990 Clean Air Act?

19. *Regulations Problem 19.* (air pollution, air quality standards, ozone). [wma]. A metropolitan area has recently measured elevated concentrations of ozone, and been declared "non-attainment" with respect to the National Ambient Air Quality Standard for this pollutant. How are each of the following types of local industries likely to be affected by this designation? Be specific.

    a.   coal-fired power plant
    b.   petroleum refinery
    c.   sawmill

20. *Regulations Problem 20.* (air pollution, air pollution regulations, air pollution law). [wma]. Air pollution regulatory strategy may include the "bubble policy." Such a policy is likely to receive wide support from industrial air pollution sources. Explain what this policy is and why industry would support it.

21. *Regulations Problem 21.* (Clean Air Act, acronyms, MACT). [hb]. The Clean Air Act and the people who work with it use a large number of acronyms with dizzying frequency. Knowledge of these acronyms is an absolute necessity for communicating with professionals in the environmental management field. Provide the meaning of the following acronyms.

Each of your answers should be approximately one or two paragraphs long.

a. MACT. Explain how MACT applies to attainment areas, non-attainment areas, major stationary sources, hazardous air pollutants (Section 112 of Title III), consideration of cost, new sources, existing sources, and the issuing of permits.

b. RACT. Explain how RACT is used in an environmental management system. How does RACT apply to existing major stationary sources, National Ambient Air Quality Standards, process modifications, considerations of cost, social impacts, and alternate approaches to meeting Clean Air Act requirements.

c. CTG. Explain how CTGs apply to existing non-attainment area major stationary sources.

d. NSPS. Discuss how NSPSs are used in an environmental management system. Explain how NSPSs apply to attainment areas, non-attainment areas, major new stationary sources, modified stationary sources, specific pollutants, energy needs, consideration of costs, work practices, alternate ways of reducing regulated emissions, and the issuing of permits.

e. LAER. Describe under what circumstances LAER might be part of an environmental management system. Discuss how LAER applies to attainment areas, non-attainment areas, new major stationary sources, modified major stationary sources, specific pollutants, specific sources, categories of sources, considerations of cost, and the issuing of permits.

21. *Regulations Problem 21.* [hb] (continued)

    f. <u>PSD</u>. Discuss how PSD might be incorporated into an environmental management system. Explain how PSD applies to attainment areas, new stationary sources, the modification of stationary sources, specific sources, specific pollutants, hazardous air pollutants (Section 112 of Title III), specific categories of sources, quantities of pollutant(s) emitted, consideration of cost, considerations of energy use, economic considerations, and the issuing of permits. Indicate why the application of PSDs is controversial.

22. *Regulations Problem 22.* (BACT, MACT, specific sources, source categories). [hb]. Environmental management systems must obviously take into account environmental laws and regulations. In some cases, the job is severely complicated by disputes over the meaning of the regulations. For example, the Clean Air Act of 1990 and the regulations issued by the U.S. EPA implementing the Act speak of the Maximum Achievable Control Technology (MACT) as well as of the Best Available Control Technology (BACT). These two terms appear to have the same meaning, i.e., the "Maximum" would also have to be the "Best," and if the technology is "Achievable," it would also have to be "Available."

Explain the difference between the terms MACT and BACT. How is each term defined and applied?

III. SOLUTIONS

1. *Regulations Solution 1.* [sn].

- National Environmental Policy Act (NEPA) of 1969. Establishes a federal policy of environmental protection for and by all Americans. Requires the preparation of environmental impact statements for all federal actions. Establishes the Council on Environmental Quality.
- Clean Air Act (CAA) of 1970; last amended in 1990. Seeks to protect the nation's air quality by imposing emission standards on stationary and mobile sources of air pollution.
- Occupational Safety and Health Act of 1971. Focused on work place safety and regulates safe use and disposal of chemicals. Information developed by the National Institutes of Occupational Health (NIOSH) under this law helps in assessing health risks of chemicals in the environment. Material Safety Data Sheets (MSDSs) produced as part of the "Worker-Right-To-Know" provisions of OSHA are used to comply with the "Community-Right-To-Know" requirements under CERCLA (see below).
- Federal Water Pollution Control Act Amendments of 1972. (Name changed to the Clean Water Act (CWA) by 1977 amendments). Seeks to restore and maintain the physical, chemical, and biological integrity of the nation's waters by imposing the National Pollutant Discharge Elimination System (NPDES) permit system, setting in-stream water quality standards, and requiring clean up of oil spills.
- Comprehensive Environmental Response, Compensation and Liability Act (CERCLA) of 1980, commonly known as "Superfund." Attempts to clean up abandoned hazardous waste sites that were in existence prior to the passage of RCRA (see below).
- Federal Insecticide, Fungicide and Rodenticide Act (FIFRA) of 1947. Seeks to ensure that society reaps the benefits of pesticide use, with minimum risk to human health and the environment.
- Resource Conservation and Recovery Act (RCRA) of 1976. Requires "cradle to grave" management of hazardous waste including permitting of hazardous waste treatment, storage, and disposal facilities. Also establishes rules governing the proper disposal of non-hazardous solid waste.
- Hazardous and Solid Waste Amendments (HSWA) of 1984. Broadens the scope of RCRA and includes provisions to protect the quality of groundwater.
- Pollution Prevention Act (PPA) of 1990. Mandates a national policy regarding a hierarchy of preferred waste management approaches: source reduction, recycle, treatment, and disposal, all to be conducted in an environ-

1. *Regulations Solution 1.* [sn] (continued)

mentally sound manner. The act also encourages industry to voluntarily reduce the amount of waste generated during the manufacturing process.

- Safe Drinking Water Act (SDWA) of 1974, last amended in 1996. Aims for the protection of "sole source" aquifers and other aquifers from contamination resulting from underground injection of waste.
- Superfund Amendments and Reauthorization Act (SARA) of 1986. Addresses concepts regarding financial liability for hazardous waste. Established the "Community-Right-To-Know" rules under Title III, requiring industrial disclosure of legal releases of hazardous substances to the environment, and the development of state and local planning committees organized to prepare for emergency responses to accidental chemical releases.
- Toxic Substance Control Act (TSCA) of 1976. Regulates the manufacturing, processing, distribution in commerce, use, and disposal of certain chemical substances and mixtures.

2. *Regulations Solution 2.* [dls]. Some environmental laws make reference to other laws and recent regulations draw authority from more than one law to accomplish an environmental management goal. The courts use precedence from cases tried under one environmental law to support a decision under another law. In general, regulatory agencies and the courts tend to use their authority and the regulations from any source in their jurisdiction to protect human health and the environment. Even tax and business regulations may be used in enforcement actions, where appropriate, to force companies to do what is prudent in order to protect the environment.

A prominent example of this concept is the National Oil and Hazardous Substances Pollution Contingency Plan (40 CFR § 9 and 300) also known as the National Contingency Plan or NCP. This regulation, which was developed by U.S. EPA to establish a system for responding to oil and hazardous substances releases, draws its authority from both the Clean Water Act and the Comprehensive Environmental Response, Compensation and Liability Act (Superfund). The NCP regulations are multimedia in scope, i.e., they deal with pollution of water, soil, and the atmosphere.

Another example lies within Superfund itself. This law requires that any substance listed as hazardous under any other environmental law (e.g., Clean Air Act, Clean Water Act, Safe Drinking Water Act, etc.) is to be considered as hazardous under Superfund.

3. *Regulations Solution 3.* [dls]. Common Law represents the rules and principles relating to the government and security of

3. *Regulations Solution 3.* [dls] (continued)

persons and property that is derived from natural reason, our innate sense of justice, and the dictates of conscience. Its authority is derived from uses and customs recognized and enforced by the courts. Common Law deals with complaints brought to court because of negligence, trespass, or nuisance. In theory, any citizen with a legitimate complaint can sue another citizen or company who has wronged him or her in order to force them to change their behavior and pay for damages suffered because they were negligent, etc. There does not have to be any violation of law by the person or company being sued. This can be a powerful tool against polluters when their contamination of the air, surface water, groundwater, or soil environment impacts neighbors.

4. *Regulations Solution 4.* [dls]. People in the western world began to recognize the role of a clean environment in protecting their health with the substantiation of the germ theory of disease in the latter half of the 19th century. The first federal environmental law (The River and Harbors Act) was passed in 1899. In 1912, the U.S. Public Health Service was established. From about this point in time, the incidence in the U.S. of many diseases has decreased and life expectancy has increased. Of course, major advances in food preservation and medical practice have also contributed substantially to improved health, but many of the improvements are due to a cleaner environment, including cleaner water and air. These improvements have paralleled the passage of federal environmental laws and the changes in environmental management that have come about because of the requirements of these laws.

5. *Regulations Solution 5.* [dls]. Possible Answer 1. Yes. The U.S. system of environmental management evolved in response to public perceptions of a need to provide a cleaner environment following the industrialization of this country. This system is, in general, able to protect public health in a complex, industrial society. Developed and developing nations of the world who choose to model their environmental management system after that of the U.S. take advantage of decades of experience in addressing environmental problems.

Possible Answer 2. No. While the U.S. has a highly developed and relatively effective system of managing the environment and protecting public health in an industrialized nation, its system is dependent on the prosperity and public service infrastructure that exists in this country. The costs of environmental protection in the U.S. have been excessive, i.e., the system has not been cost-effective. For most of the world, neither the financial freedom, nor the public infrastructure exists to support an environmental

5. *Regulations Solution 5.* [dls] (continued)

management system like that currently used in the U.S. Other systems must be developed that may use basic ideas from the U.S. system, but which provide other incentives and recognize the customs and limited monetary resources of the nations where they will be applied.

6. *Regulations Solution 6.* [dls]. The Fifth Amendment to the Constitution requires that citizens be compensated fairly for property taken by the government for the public good. It has been argued that environmental regulations that interfere with the use of property reduce the value of the property, effectively "taking" it from the owner. Wetlands protection provisions under the Clean Water Act and habitat protection requirements under the Endangered Species Act are the most common causes for this issue to arise.

7. *Regulations Solution 7.* [dls]. The National Oil and Hazardous Substances Pollution Contingency Plan (40 CFR § 9 and 300), also known as the National Contingency Plan (NCP), has a requirement under CERCLA remediation rules to:

> "Establish remedial action objectives specifying contaminants and media of concern, potential exposure pathways, and remediation goals. Initially, preliminary remediation goals are developed based on readily available information, such as chemical-specific ARARs..."

ARARS are <u>Applicable or Relevant and Appropriate Requirements</u> and are defined as, "Those cleanup standards, standards of control, and other substantive requirements, criteria, or limitations promulgated under federal environmental or state environmental or facility siting laws that, while not 'applicable' to hazardous substance, pollutant, contaminant, remedial action, location, or other circumstances at a CERCLA site, address problems or situations sufficiently similar to those encountered at the CERCLA site that their use is well suited to the particular site. Only those state standards that are identified in a timely manner and are not more stringent than federal requirements may be relevant and appropriate."

8. *Regulations Solution 8.* [dls]. Many companies and government agencies had pollution prevention programs in operation prior to 1990. 3M calls their program Pollution Prevention Pays (3Ps). The 3Ps program contains provisions for recognizing 3M employees who suggest changes in operations that result in less pollution by the company. The Pollution Prevention Act requires U.S. EPA to establish an annual award program to

8. *Regulations Solution 8.* [dls] (continued)

recognize companies which operate outstanding or innovative source reduction programs. It also requires U.S. EPA to establish a system for sharing pollution prevention information and technology among U.S. companies.

9. *Regulations Solution 9.* [hb]. The six surrogate enforcers are:
   a. banks
   b. insurance companies
   c. accounting firms
   d. investors and shareholders
   e. international standards
   f. the public

a. Banks regularly lend large amounts of money to corporations to enable them to conduct business and to grow. Indeed, large corporations could not function in the modern economy without such loans. Even small companies routinely borrow money from banks. The banks hope, of course, to make a profit from the interest on the loans. Of vital concern to the bank, therefore, is the ability of the corporation or small business to repay the loans. Banks have become acutely aware that environmental liabilities could reduce a company's ability to repay a loan and even drive large corporations (such as Johns-Manville) into bankruptcy. Should a bank have to foreclose on a company, it might even inherit the company's environmental liabilities. Banks thus check into a company's environmental liabilities which has the indirect effect of enforcing environmental regulations.

b. Insurance companies provide many types of insurance policies to protect companies from a wide variety of risks. Among them are policies that insure against environmental risk and protect the officers of the company against direct liability claims. Insurance companies base premiums on their estimate of the risk of having to pay a benefit. To determine that premium, insurance companies investigate a company's environmental liabilities. Poor environmental management practices would result in high premiums or perhaps inability to get any insurance at all. Such an outcome would, at the least, reduce the profits of the company and could prevent it from getting the investments it needs to continue in business. The officers of the company would also feel directly and personally threatened if they could not get insurance to protect them from environmental liability generated law suits. The company is thus motivated by the insurance companies to adopt sound environmental management practices. Consequently, the insurance companies are indirectly enforcing environmental regulations.

9. *Regulations Solution 9.* [hb] (continued)

c. Recent actions by the Securities and Exchange Commission (SEC) have required accounting firms to clearly report a corporation's environmental liabilities on the balance sheet of the corporation's required audit. Poor environmental management practices leading to significant liabilities will now be very difficult to hide. The SEC has threatened to suspend trading in the stock of any corporation that does not comply with the new rules. Accounting firms that do not comply could risk their very existence. Thus, the SEC and accounting firms indirectly enforce environmental regulations

d. The actions of banks, insurance companies, and accounting firms based on a corporation's environmental management practices and consequent environmental liabilities are more visible than ever to investors and shareholders. Investors will avoid companies with such liabilities and shareholders might sell or, worse, sue the officers of the corporation. The threat of falling stock prices and law suits again result, indirectly, in enforcement of environmental regulations.

e. ISO 14000 is a voluntary Standard of Environmental Management Systems developed in Geneva, Switzerland by the International Organization for Standardization (ISO). It does not require compliance with the regulations of the country in which the company is located. If, as expected, many countries adopt laws that require imported products to have been produced by companies certified as adhering to ISO 14000, then environmental practices will almost certainly be improved. Although compliance with the regulations is not a requirement for ISO 14000 certification, good environmental management systems are. It seems likely that an increase in good environmental management systems would also result in increased compliance with the regulations. The voluntary ISO 14000 standard is thus indirectly resulting in enforcement of environmental regulations.

f. The public has demonstrated repeatedly that it supports good environmental management practices. The public is more aware than ever of environmental issues and of the companies that have a poor environmental image. A significant portion of the public is even willing to pay more for a product that is produced in an environmentally friendly way than a cheaper product from an environmentally unfriendly producer. Corporations listen closely to the desires of the buying public. The result is that the public is yet another indirect enforcer of environmental regulations.

10. *Regulations Solution 10.* [mh]. The acronym OSHA stands for the Occupational Safety and Health Act and the Occupational Safety and Health Administration which was formed under the provisions of the Act. OSHA was created in 1970, the same year that the U.S. EPA was formed. OSHA was created to protect safety

10. *Regulations Solution 9.* [mh] (continued)

and health in the workplace. Along with the EPA acting as an overlapping environmental organization, OSHA is mandated to reduce the exposure of hazardous substances over land, sea, and air. OSHA requires employers to communicate to employees on hazardous materials that exist in the workplace. It is important to note that OSHA's jurisdiction is restricted to the workplace.

11. *Regulations Solution 11.* [dj]. As defined in Section 1004-Definitions, of the Hazardous and Solid Waste Amendments of 1984 (Public Law 98-616, commonly known by the acronym HSWA) to the Solid Waste Act of 1965, a waste or combination of wastes is "hazardous" if its quantity, concentration, or physical, chemical or infectious characteristics may either: (a) cause, or significantly contribute to an increase in mortality or an increase in serious irreversible or incapacitating reversible illness, or (b) pose a substantial present or potential hazard to human health or the environment when improperly treated, stored, transported, disposed of, or otherwise managed (Holmes et al., 1993).

Holmes, G., Singh, B. and Theodore, L., Environmental risk assessment, in *Handbook of Environmental Management and Technology*, Wiley-Interscience, New York, 1993, p. 365.

12. *Regulations Solution 12.* [dj]. Under 40 CFR 261.21 through 261.24, a waste is hazardous if it has any one or more of the following properties:

1. Corrosivity, which is defined in 40 CFR as having a pH < 2.0 or pH > 12.5
2. Ignitability, defined as any waste with a flash point less than 140°F (60°C).
3. Reactivity. A waste is reactive if it exhibits any of the following characteristics:
   a. It undergoes a violent change with or without detonating.
   b. It undergoes violent reaction with water or forms unstable or dangerous mixtures with water.
   c. It is a cyanide-bearing or sulfide-bearing waste capable of releasing toxic gases, vapors, or fumes at pH's between 2.0 and 12.5 in such quantity as to cause harm to human health or the environment.
4. Toxicity. A waste is toxic if, after undergoing one of two specified leaching procedures, the leachate contains any one of a set of listed chemicals at a concentration above a given threshold value. In a leaching procedure, a specified quantity of waste is extracted in a known volume of acidic solution for a specified period of time.

12. *Regulations Solution 12.* [dj] (continued)

> The resulting "soup" of water and substances that leached from the waste is then assayed for the concentrations of a set of harmful chemicals that are known to be very toxic to animal and plant life. There are two leaching procedures in use, the EP (Extraction Procedure) test, and the TCLP (Toxic Compound Leaching Procedure) test. As an example, after undergoing a TCLP procedure, a waste's leachate contains 3.0 mg/L of cadmium. Since this value exceeds the specified 1.0 mg/L threshold value for cadmium in TCLP leachate, the waste is classified as hazardous.

13. *Regulations Solution 13.* [dj].

a. The acronym CERCLA stands for Comprehensive Environmental Response, Compensation and Liability Act.

b. CERCLA was enacted in 1980. Congress passed the legislation to respond to the risks to public health and property that had been caused by prior practices of the disposal of hazardous wastes. The legislation was intended to compel the clean up of abandoned, uncontrolled hazardous waste sites.

c. Superfund is a trust fund that was set up allowing the U.S. government to pay for the clean up of a hazardous waste site. In 1980 the initial $1.6 billion in Superfund came from a tax on commercial chemicals and crude oil. Currently, moneys come from a combination of a tax on petrochemicals and from general revenues. Superfund can be used to pay for the investigation, clean up and closure of a hazardous waste site before a responsible party can be located to pay the costs. If a responsible party or parties can be found, then the U.S. EPA can require them to reimburse Superfund for the cost of the clean up. In cases where no responsible party can be found, Superfund is not reimbursed for the clean up.

The intent of Superfund, and other parts of CERCLA, was to compel all concerned parties to "clean it now, argue later," so that remediation of current threats to public health would not be delayed by years or decades of litigation.

14. *Regulations Solution 14.* [mh]. Superfund usually seeks voluntary compliance. If an agreement is not found a suit naturally follows. The responsible party could be charged three times what it cost the government to clean up the site. The three main powers the EPA possesses are listed below:

1. Ex Post Facto: A party may be liable for what was once legal, but now is illegal.

14. *Regulations Solution 14.* [mh] (continued)

2. Innocent landowner liability: Anyone who owned the site may be liable for costs even if they did not know the site was contaminated.
3. Joint and several liability: Costs may be shared or picked up by just one party. This regulation allows the EPA to collect from just one party, rather than sue each individual land owner. In return, the sued alleged violator may sue the other owners for damages.

15. *Regulations Solution 15.* [wma].

a. Comprehensive Environmental Response, Compensation & Liability Act (CERCLA/Superfund), and Resource Conservation and Recovery Act (RCRA)

b. Contact the appropriate local environmental or public health agency. Have the unknown waste analyzed to determine whether it is a hazardous or non-hazardous waste. Analysis will utilize the tests for hazardous characteristics:

- Corrosivity
- Ignitability
- Reactivity
- Toxicity

c. Dispose of material at a licensed facility, utilizing a Uniform Hazardous Waste Manifest if hazardous.

16. *Regulations Solution 16.* [hb].

a. Toxic Release Inventory (TRI) reporting requirements cover facilities which meet the following conditions:

- the facility manufactured or imported or processed in excess of 25,000 pounds any of the 302 individual chemicals and 20 chemical categories listed for required TRI reporting or they used in any other manner 10,000 pounds or more of a TRI chemical during the calendar year;
- the facility was engaged in general manufacturing activities (meaning that they were classified within Standard Industrial Classification (SIC) Codes 20 through 39); and
- the facility employed the equivalent of ten or more full-time employees.

b. Toxic Release Inventory reporting requirements call for facilities to report the amounts of the listed toxic chemicals that are

16. *Regulations Solution 16*. [hb] (continued)

released directly to air, water, land, or injected in underground wells. Facilities must also report amounts of chemicals that are transported or transferred off-site to facilities that treat, store, or dispose of the chemical wastes. Releases and transfers include the following:

- emissions of materials to the air;
- wastewater discharges into rivers, streams, or other bodies of water;
- releases to land on site, including landfill, surface impoundment, land treatment, or any other mode of land disposal;
- disposal of wastes in underground injection wells;
- transfers of wastewater to publicly owned treatment works (POTWs); and
- transfers of wastes to other off-site facilities for treatment, storage, or disposal.

Both routine releases and accidental spills or leaks must be reported. Facilities must report the release even if their releases comply with all environmental laws and permits.

17. *Regulations Solution 17*. [mh]. Litigation is the number one reason why so few sites have been cleaned up. These legal actions have generated 50 to 90% of all expenditures in the Superfund program, resulting in only 10 to 50% of the expenditures on actual clean up costs. In addition, states and communities are fighting over which sites shall be cleaned first. Finally, as is the case in most bureaucratic programs, each Superfund site must be carried through a specific, exhaustive, ten-step procedure regardless of whether each of the ten steps is necessary or not. Only after these ten steps have been completed may corrective action officially begin at a Superfund site.

18. *Regulations Solution 18*. [mh]. The 1990 Clean Air Act set new requirements for smaller sources (referred to as area sources) of air pollution. State and local governments will ultimately be responsible for management and enforcement of such new requirements. State and local governments must also develop technical assistance programs to help small businesses comply with these new rules.

The areas of the 1990 Clean Air Act where its goals and approaches will impact small businesses include:

- Photochemical smog (affects the release of volatile organic compounds, principally hydrocarbons)
- Inadvertent release of gaseous air toxics

18. *Regulations Solution 18.* [mh] (continued)

- Ozone depleting materials (as listed by the EPA) by the Year 2000
- Fleet controls on those companies that have more than 10 vehicles
- New environmental permits
- Involvement with state and local governments for rules regarding compliance standards.

19. *Regulations Solution 19.* [wma].

a. Coal-fired power plants typically emit large quantities of particulate matter, sulfur dioxide, and nitrogen oxides. Since nitrogen oxides are important precursors to the formation of tropospheric ozone, the power plant can expect increased regulation to reduce emissions of this family of air pollutants.

b. Petroleum refineries typically emit both nitrogen oxides and reactive hydrocarbons. Since these are the principal pollutants contributing to the formation of ozone in the atmosphere (in the presence of sunlight), the petroleum refinery should expect more rigorous (lower) emission limits for both pollutants.

c. Sawmills typically emit particulate matter which is not a significant contributor to ozone formation. Consequently, no direct regulatory impact is expected in this industry. (Note: if the sawmill has its own heating or power source as some do, nitrogen oxides may be emitted and regulated from that part of the operation.)

20. *Regulations Solution 20.* [wma]. The bubble policy allows an industrial facility to comply with emission limits on a facility-wide basis (all sources in one "bubble"), rather than having to meet separate emission limits applied to each source operation, stack or vent which perhaps number several hundred at a large manufacturing facility! The total allowable emissions remain the same, but the facility-wide approach enables the industry to reduce emissions more cost-effectively. For example, "over controlling" or even closing some source operations, while "under controlling" others, is likely to be preferred to uniform controls on all sources.

21. *Regulations Solution 21.* [hb]. The reader will need an appropriate text book or copy of the Clean Air Act of 1990 and Title 40 of the Code of Federal Regulations (40 CFR) to answer the questions adequately.

a. <u>MACT</u>. Maximum Achievable Control Technology. MACT must be applied in both attainment and non-attainment areas to categories of major stationary sources of any of the 189 hazardous air pollutants (HAPs) listed in Section 112 of Title III of the Clean

21. *Regulations Solution 21.* [hb] (continued)

Air Act of 1990. These 189 HAPs are also known as air toxics. Cost may be considered in the development of MACTs. MACT applies to both new and existing sources but is more stringent for new sources. For new sources, MACT must "... not be less stringent than the emission control that is achieved in practice by the best controlled similar source..." For existing sources, MACT must be set at not less than the level of control achieved by the 12% of sources in the same category with the best record of performance. If a category has fewer than 30 sources, then the best five are used. MACT could be part of a permit.

    b. RACT. Reasonably Available Control Technology. RACT applies to categories of existing, non-attainment area, major stationary sources of pollutants for which the following National Ambient Air Quality Standards (NAAQS) have been set: the two ozone precursors, volatile organic compounds (VOCs) and nitrogen oxides (NO_X); carbon monoxide (CO); particulate matter smaller than 10 μm (PM-10); sulfur oxides (SO_X); and lead. RACT means devices or system process modifications, etc., that are reasonably available. RACT development takes into account the need to meet the NAAQS, cost, social and environmental impact, alternate ways of meeting the NAAQS and other factors.

    c. CTGs. Control Technique Guidelines. CTGs are documents published by the U.S. EPA detailing what constitutes RACT for each category of existing, non-attainment areas, major stationary sources of VOCs, NO_X, CO and/or PM-10. EPA has published about 30 CTGs thus far; most of them deal with VOCs. Another dozen CTGs are currently being worked on by the EPA.

    d. NSPS. New Source Performance Standards. NSPSs are emission standards for categories of major new or significantly modified stationary sources of emissions of VOCs, NO_X, CO, PM-10, acid mist, SO_X, fluorides, H_2S in acid gas, lead, and total reduced sulfur. Development of NSPSs utilizes the "emission reduction achievable" by the best technology that has been "adequately demonstrated." Environmental impact and energy needs are also considered. Cost is considered in a limited way.

NSPSs are sometimes stated as process or work practices, operational standards, design standards or other alternate ways of producing low emissions. NSPSs also serve as the minimum standards used in establishing permits for the construction of specific major new stationary emission sources. NSPSs are periodically revised to take advantage of new technology to achieve still lower emissions. NSPSs apply in both attainment and non-attainment areas. In attainment areas, the performance standard is known as the Best Available Control Technology (BACT). In non-attainment areas, the performance standard is known as the Lowest Achievable Emission Rate (LAER).

21. *Regulations Solution 21.* [hb] (continued)

e. LAER. Lowest Achievable Emission Rate. LAER applies to emission standards in permits for construction of new or significantly modified major stationary specific sources of criteria pollutants in a non-attainment area. The criteria pollutants are CO, NO$_X$, SO$_2$, lead, total particulate matter, PM-10, and ozone (and its precursor, VOCs). LAERs are applied to specific sources but are established by category of source. LAER is either: the lowest rate that any source in that category or a similar category has achieved, or the lowest emission rate in any State Implementation Plan (SIP), whichever is lower. LAER must be at least as low as any NSPS for that source category. Cost is only considered if it is so high as to render use of the LAER standard not feasible for that specific source.

f. PSD. Prevention of Significant Deterioration. To prevent areas that are in compliance with air quality standards from deteriorating in quality as the result of the construction of new sources or major modifications to existing sources of pollutants, permits are required before such construction or modification can take place. These permits apply to new specific stationary sources that could emit 100 tons per year (T/yr) or more of any regulated pollutant (except the 189 HAPs) and belongs to one of the 28 categories specifically listed by the U.S. EPA or could emit 250 T/yr or more of a pollutant. Permits are also required if a major modification to an existing plant results in an increase in emissions of any of the following pollutants in the amount given in parentheses: CO (100 T/yr), NO$_X$ (40 T/yr), SO$_2$ (40 T/yr), total particulate matter (25 T/yr), PM-10 (15 T/yr), ozone (40 T/yr of VOCs), lead (0.6 T/yr), asbestos (0.007 T/yr), beryllium (0.0004 T/yr), mercury (0.1 T/yr), vinyl chloride (1 T/yr), fluorides (3 T/yr), sulfuric acid mist (7 T/yr), H$_2$S (10 T/yr), total reduced sulfur including H$_2$S (10 T/yr), and reduced sulfur compounds including HS$^-$ (10 T/yr).

The permits require the use of the Best Available Control Technology (BACT). The development of the BACT for each specific source permit takes into account many factors on a case-by-case basis. Some of these factors are: technology, cost, energy use, and the need to limit how much any one plant can be allowed to deteriorate the air quality so that other plants needed for economic development in the area can be built. The BACT is determined by the state within rules issued by the U.S. EPA. The development of BACTs has been and still is the subject of controversy and litigation.

22. *Regulations Solution 22.* [hb]. In terms of degree of pollutant emission reduction and the means by which the reduction is achieved, MACT and BACT have much in common. MACT calls for the "maximum degree of reduction in emissions" that is

22. *Regulations Solution 22.* [hb] (continued)

achievable. BACT calls for an "emission limitation based on the maximum degree of reduction of a pollutant." In establishing an MACT standard, cost, environmental impacts and energy requirements are considered. In establishing a BACT standard, here, too, cost, environmental and economic impacts and energy requirements are considered. In addition to emission controls, MACT standards may include changes in processes, methods, design, operational changes, work practices, etc. BACT standards, in addition to emissions controls, may also include changes in production processes, methods, systems, etc.

The terms are applied, however, in different ways. MACT applies in both attainment and non-attainment areas to categories of major new and existing stationary sources of the 189 HAPs listed in Title III and known as toxic air pollutants. BACT applies only in attainment areas and to specific new or significantly modified existing stationary sources of all pollutants regulated under the Clean Air Act of 1990 except the 189 HAPs listed in Title III.

Determining the allowable emission levels is more straight-forward under MACT than BACT. The MACT is different for new versus existing sources. For new sources, MACT must "...not be less stringent than the emission control that is achieved in practice by the best controlled similar source..." For existing sources, MACT must be set at not less than the level of control achieved by the 12% of sources in the same category with the best record of performance. If a category has less than 30 sources, then the best five are used.

The BACT is determined on a case-by-case basis by the state under rules issued by the U.S. EPA. The state considers a host of factors when setting the BACT including politically sensitive issues such as the amount of economic development to be allowed in the area and by how much any one source should be allowed to degrade clean air. The EPA, however, has issued rules that, in effect, require the most stringent emission controls and that minimize cost considerations. This has led to much controversy and even lawsuits. The issue remains unresolved.

Chapter 2

HEALTH AND HAZARD RISK ASSESSMENT

Scott Lowe and David James

I. INTRODUCTION

A. EXAMPLES OF RISK ASSESSMENT

Assessments of risk can be simple or complex, depending on the size and scale of the problem. For example, risk assessment for the accidental spill of battery acid in an auto repair shop might include:

1. Obtaining data on the chemical and toxicological properties of sulfuric acid;
2. Training personnel in the proper procedures and precautions for handling automobile batteries;
3. Training personnel in the proper methods for clean up of an acid spill;
4. Having protective gear available for personnel who are designated to clean up any spill; and
5. Having spill clean-up equipment available in a prominent and unobstructed location in the shop.

In contrast, the risk assessment for a petrochemical complex located on the Gulf of Mexico would involve assessment of dozens of different processes and, potentially, hundreds of different compounds. Such a facility might very well have a team of full-time specialists who continually assess risks, request maintenance of suspect equipment, and conduct training sessions for all the plant workers. Contingency plans for evacuation of the complex and the surrounding community would usually have to be in place, as well as plans for access to the complex by firefighters, police and emergency medical personnel. The hazard risk assessment, including maintenance and inspection procedures, training and response plans could require hundreds of printed pages, the investment of a sizable sum of money, and the participation and approval of several government agencies.

B. EXAMPLE OF A HEALTH RISK ASSESSMENT

The residents of a small town rely on a community well system for their drinking water. The wells are monitored every 3 months for the presence of contaminants. The most recent quarterly monitoring has revealed the presence of chlorinated solvents at concentrations near the federal maximum contaminant levels (MCLs). Further investigation reveals that the solvents are coming from the town's landfill, where cleaning solvents from a small agricultural implement company were disposed of over a period of three decades. These solvents have migrated into the drinking water aquifer, and a sizable volume of groundwater has been contaminated. The public becomes extremely concerned and demands that the affected wells either be shut down or treated. The state instructs the water utility and town council to conduct a health risk assessment to determine if the concentration and duration of exposure have been sufficient to impair human health. Lifetime cancer risk becomes an important issue, and the town council decides to request a risk assessment from the state Department of Environmental Quality (DEQ).

The state DEQ must then estimate the lifetime cancer risk that results from drinking groundwater contaminated with trace amounts of chlorinated solvents. After the health risk assessment is completed for the hypothetical situation presented above, it is the responsibility of the participating stakeholders (including state and local elected officials and the public) to decide on a course of action. Action can include doing nothing (if the risk is low, and if "no action" will be politically acceptable), capping the affected wells and providing residents with an alternative water supply, remediating the contaminated aquifer to prevent entry of contaminated water into the distribution system, or treating the contaminated water as it enters the distribution system. A comprehensive risk assessment might also include evaluating the health risks associated with each possible course of action.

C. EXAMPLE OF A HAZARD RISK ASSESSMENT

In contrast to the chronic (long-term) exposures that predominate in health risk assessments, acute (short-term) exposures are of major concern in hazard risk assessments. In May 1991 an accidental release of chlorine gas from a broken fitting at the Pioneer Chlor-Alkali plant in Henderson, Nevada released a gas cloud that covered about 1 square mile. The incident resulted in the evacuation of several thousand people from their homes and their relocation to emergency shelters. Respiratory distress was documented in several hundred people, schools, factories and shopping areas were shut down, and transportation in the area was disrupted for more than 12 hours. The factory that was responsible for the leak paid a large fine and settled numerous civil damage claims caused by the release.

Fortunately, no one was killed as a result of exposure to the cloud. Chlorine leaked out at a fairly slow rate and winds were light, giving people time to either get out of the path of the plume or to implement shelter-in-place procedures. Trained emergency response teams were able to cordon off the affected area and assist in the notification and evacuation of persons from areas in the projected path of the chlorine plume. The accident occurred in the early morning hours, and, once the leak had been fixed, the plume was diluted by atmospheric mixing. The plume had dissipated to safe levels by nightfall.

The reader may have seen televised accounts or read of this incident or similar events which have occurred in their local area. In addition to industrial accidents, transportation accidents, such as train derailments or the overturning of a tanker truck, can also result in the release of hazardous materials into the environment if these accidents result in the rupture of chemical containers.

When properly contained and handled, industrial chemicals can safely serve the needs of a modern industrialized society. However, some of these materials can pose a significant threat to health, life, or property if they are accidentally released in large quantities.

Under guidelines established by the Resource Conservation and Recovery Act (RCRA) of 1984, wastes are defined to be hazardous if they are corrosive, reactive, flammable, or toxic. In addition to the general definitions, the federal government has designated many specific compounds (such as chloroform) and wastes from specific types of industries (such as spent halogenated solvents used in degreasing) as hazardous.

The Hazardous Materials Transportation Act of 1975 classifies hazardous materials that are subject to federal transportation regulations. Hazardous materials must be properly manifested, packaged, labeled, and externally placarded when being shipped by any mode of transportation.

Hazard risk assessment is the process of assessing the likelihood and severity of a hazardous material release from a particular location, developing designs and procedures to minimize the chance of a release, and developing plans for response to a release if one should occur. Proper planning, containment, prevention, and emergency response used together minimize both the likelihood and risk of exposure of an accidental chemical release.

D. ABOUT THE PROBLEMS IN THIS CHAPTER

The problems in this chapter are intended to introduce the reader to the legislation governing risk assessments and the major steps involved in the risk assessment process. Many focus on information needs, calculations, and judgments that are part of a health risk assessment. In addition, the types of chemical emergencies and the major features of emergency preparedness plans are also covered. Communication of risk, an integral part of the risk assessment process, is covered in Chapter 3.

REFERENCES

1. **Holmes, G., Singh, B. and Theodore, L.**, Environmental risk assessment, in *Handbook of Environmental Management and Technology*, Wiley-Interscience, New York, 1993, chap. 33.
2. **LaGrega, M., Buckingham, P. and Evan, J.**, Quantitative risk assessment, in *Hazardous Waste Management*, McGraw-Hill, New York, 1994, chap. 14.
3. **Masters, G.**, Hazardous substances and risk analysis, in *Introduction to Environmental Engineering and Science*, Prentice-Hall, New York, 1991, chap. 5.
4. **Wentz, C.**, Toxicology and risk assessment, in *Hazardous Waste Management*, 2nd Ed., McGraw-Hill, New York, 1995, chap. 2.
5. **Wentz, C.**, Transportation and storage of hazardous wastes, in *Hazardous Waste Management*, 2nd Ed., McGraw-Hill, New York, 1995, chap. 10.

II. PROBLEMS

1. *Health and Hazard Risk Assessment Problem 1.* (OSHA, employee safety, employer responsibility). [rt]. The Occupational Safety and Health Act (OSHA) enforces basic duties which must be carried out by employers. Discuss these basic duties

2. *Health and Hazard Risk Assessment Problem 2.* (OSHA, NIOSH, industrial exposure, worker safety). [rt]. State the major roles of the National Institutes of Safety and Health (NIOSH) and the Occupational Safety and Health Administration (OSHA).

3. *Health and Hazard Risk Assessment Problem 3.* (health risk assessment, exposure pathway) [pcy]. List and describe four major steps in a health risk assessment.

4. *Health and Hazard Risk Assessment Problem 4.* (health risk assessment, risk communication, risk perception, risk estimation, comparative risk assessment). [pcy]. Define the following terms:

   a. Risk management
   b. Risk communication
   c. Risk estimation
   d. Risk perception
   e. Comparative risk assessment

5. *Health and Hazard Risk Assessment Problem 5.* (risk assessment, risk management). [dls]. What is the role of human health risk assessment in risk management?

6. *Health and Hazard Risk Assessment Problem 6.* (reference dose, dose-response curve). [dls]. Describe and illustrate the process of setting a reference dose (RfD) using a schematic dose-response curve. Correctly label the axes and all other important information on your illustration.

7. *Health and Hazard Risk Assessment Problem 7.* (exposure assessment, health risk assessment, dose-response). [dls]. Fill in the following blanks.

   For the purposes of an exposure assessment done as part of a risk assessment, _____ is defined as the contact of a chemical, physical or biological agent with the _____ boundary of an organism. It is quantified as the _____ of the agent in the medium contacted, integrated over the _____ of the contact.

8. *Health and Hazard Risk Assessment Problem 8.* (average person). [sxl]. For risk assessments applied to large groups of individuals certain assumptions are usually made about an "average" person's attributes. List the "average" or standard values used for:

    a. Body weight
    b. Daily drinking water intake
    c. Amount of air breathed per day
    d. Expected life span
    e. Dermal contact area

9. *Health and Hazard Risk Assessment Problem 9.* (acute effects, averaging time, carcinogenic effects, health risk assessment). [dls]. What averaging times are typically used for evaluating acute effects versus carcinogenic effects in exposure quantification?

10. *Health and Hazard Risk Assessment Problem 10.* (health risk assessment, modeling, exposure assessment). [dls]. Why is predictive mathematical modeling used in the exposure assessment process in health risk assessments?

11. *Health and Hazard Risk Assessment Problem 11.* (health risk assessment, exposure assessment, uncertainty). [dls]. Human health risk assessments are generally not considered to be precise. Much of the uncertainty in the assessments can be associated with the exposure assessment step.

Give a correct definition of uncertainty as applied to exposure assessments and describe the sources of some of this uncertainty.

12. *Health and Hazard Risk Assessment Problem 12.* (health risk assessment, risk uncertainty). [wma]. Calculating the risk to human health attributable to an environmental contaminant often begins with the following simple equation:

Health Risk = (Human Exposure) (Potency of Contaminant)

Identify several factors that may lead to an "uncertain" calculation of the actual risk to human health of a particular chemical.

13. *Health and Hazard Risk Assessment Problem 13.* (health risk assessment, fate and transport analysis). [dls]. List at least four questions that should be answered in the evaluation of the fate and transport of a released contaminant.

14. *Health and Hazard Risk Assessment Problem 14.* (drinking water, MCL, atrazine, reference dose). [dls]. The drinking water maximum contaminant level (MCL) set by the U.S. EPA for atrazine is 0.003 mg/L and its Reference Dose (RfD) is 3.5 mg/kg-d. How many liters of water containing atrazine at its MCL would a person have to drink each day to exceed the RfD for this triazine herbicide?

15. *Health and Hazard Risk Assessment Problem 15.* (health risk assessment, dose-response, exposure assessment, risk characterization). [dls]. The diagram below illustrates the relationships among the four basic elements of human health risk assessment. Fill in the blank labels to complete the figure.

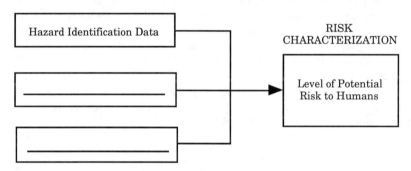

Figure 1. Relationship among basic elements of a human health risk assessment.

16. *Health and Hazard Risk Assessment Problem 16.* (health risk assessment, exposure assessment, Monte Carlo simulation). [dls]. What is Monte Carlo simulation and how is it used in risk assessment practice?

17. *Health and Hazard Risk Assessment Problem 17.* (health risk assessment, dose-response, unit risk). [dls]. What is the definition of carcinogenic "unit risk"?

18. *Health and Hazard Risk Assessment Problem 18.* (health risk assessment, benzene, unit risk, cumulative cancer risk). [dls]. The odor perception threshold for benzene in water is 2 mg/L. The benzene drinking water unit risk is $8.3 \times 10^{-7}/(\mu g/L)$. Calculate the potential benzene intake rate (mg benzene/kg-d) and the cumulative cancer risk from drinking water with benzene concentrations at half of its odor threshold for a 30 year exposure duration.
Use the following equation for estimating the benzene ingestion rate:

18. *Health and Hazard Risk Assessment Problem 18.* [dls] (continued)

$$\text{Ingestion Rate (mg/kg-d)} = \frac{(C)\,(I)\,(EF)\,(ED)}{(BW)\,(AT)}$$

where C = concentration, mg/L; I = water ingestion rate, 2 L/d; EF = exposure frequency, 350 d/yr; ED = exposure duration; BW = body weight, 70 kg; and AT = averaging time, 70 yr.

19. *Health and Hazard Risk Assessment Problem 19.* (exposure assessment, groundwater, hydrocarbon). [dj]. A consulting engineer has been hired by a state agency to perform an exposure assessment at a private dwelling. The dwelling is located over a shallow aquifer that has been contaminated by a gasoline leak from an underground storage tank at the corner gasoline station. List at least eight important environmental factors that the consulting engineer should consider when visiting the dwelling.

20. *Health and Hazard Risk Assessment Problem 20.* (risk assessment, carcinogens). [jr]. The following equation is used in risk assessment studies for carcinogens:

$$Cm = \frac{(R)\,(W)\,(L)}{(P)\,(I)\,(A)\,(ED)}$$

where Cm =action level, i.e., the concentration of carcinogen above which remedial action should be taken; R = risk, or the probability of contracting cancer; W = body weight; L = assumed lifetime; P = potency factor; I = intake rate; A = absorption factor, the fraction of carcinogen absorbed by the human body; and ED = exposure duration.

Determine the action level in $\mu g/m^3$ for an 80 kg person with a life expectancy of 70 years exposed to benzene over a 15-year period. The "acceptable" risk is one incident of cancer per 1 million persons or 10^{-6}. Assume a breathing (intake) rate of 15 m^3/d and an absorption factor of 75%. The potency factor for benzene is 1.80 $(mg/kg-d)^{-1}$.

21. *Health and Hazard Risk Assessment Problem 21.* (risk assessment, carcinogens). [jr]. A reagent bottle containing 1 kg of benzene falls to the floor of a small enclosed storage area measuring 10 m^3 in volume. In time, the spilled liquid benzene completely evaporates. With this information answer the following questions:

    a. If 5% of the volume of benzene vapor remains trapped in the storage room, what is the benzene concentration in mg/m³?

21. *Health and Hazard Risk Assessment Problem 21.* [jr] (continued)

    b. According to the risk assessment model given in the previous problem (Health Risk Assessment Problem 20), what exposure time (in hours) would evoke a risk of 0.1 for the person described in that problem? Consider the benzene concentration found in Part a to be the "action level."

22. *Health and Hazard Risk Assessment Problem 22.* (risk assessment, toxicity, REL, dispersion, air quality models, CAPCOA, SCREEN3). [jr]. According to the 1992 CAPCOA Risk Assessment Guidelines of the State of California, the acute reference or "acceptable" exposure limit (REL) for the chemical ethyl-glycol-methyl ether is 370 $\mu g/m^3$. This means that a 1-hour exposure to a concentration of this chemical greater than 370 $\mu g/m^3$ may cause reproductive toxicity in humans.

    a. If 107 g/s of ethyl-glycol-methyl ether is being emitted from a 40-m stack (actual height), to what concentration of the chemical would a person standing 0.5 km directly downwind of the stack (i.e., under the plume centerline) be exposed? The stack inside diameter is 1.5 m. At the stack exit, the gas temperature and volumetric flow rate are 300°C and 750 acfm, respectively. The ambient air temperature is 20°C. The anemometer (to measure wind velocity and direction) is at the standard height of 10 m. The average wind velocity is 2.60 m/s and the stability category is E. The terrain is flat with no obstructions for at least 10 km in the plume direction (the stack is located in a rural area). For purposes of the model, use the Brode 2 mixing height method to determine the plume rise.

    b. According to the acute REL, is this person at risk?

    c. Determine the approximate distance (i.e., the minimum and maximum centerline distances from the stack) within which the person would be exposed to a concentration higher than the REL.

To answer the questions, use EPA's model (i.e., computer program) "SCREEN3." A copy of this model may be obtained by following the steps below:

22. *Health and Hazard Risk Assessment Problem 22.* [jr] (continued)

1. Access the web site, http://www.epa.gov/scram001, on the Internet.
2. Click the mouse on "Screening Models."
3. To download the model, click on "SCREEN3." This action will copy the model (as a ZIP file) to your computer; you will be asked to choose the directory into which the zip file (called "SCREEN3.ZIP") is to be copied.
4. To unzip (or uncondense) the file, the DOS utility program PKUNZIP.EXE should be used. If there is no such program on your computer, it can be found on the Internet by using a search engine such as YAHOO, and downloaded to your computer. To unzip the downloaded model, the following sample instruction may be typed at the DOS prompt:

   PKUNZIP SCREEN3.ZIP C\TEMP\SCREEN3.EXE

   This will unzip the file named SCREEN3.ZIP and save it as an executable file named SCREEN3.EXE in the C\TEMP directory.
5. To execute SCREEN3.EXE, simply type SCREEN3 at the DOS prompt.

23. *Health and Hazard Risk Assessment Problem 23.* (community right-to-know, hazardous chemicals, planning). [dj]. In a brief, one paragraph essay, indicate what the acronym SARA represents, describe the other name of this piece of legislation, state when it was passed, and explain why it was passed.

24. *Health and Hazard Risk Assessment Problem 24.* (community right-to-know, hazardous chemicals, planning). [dj]. List at least three specific types of chemical emergencies that might occur in a large metropolitan area.

25. *Health and Hazard Risk Assessment Problem 25.* (community right-to-know, hazardous chemicals, planning). [dj]. Briefly describe at least five specific features of an emergency preparedness plan that would be put in place to respond to a major accidental release of a volatile hazardous chemical.

III. SOLUTIONS

1. *Health and Hazard Risk Assessment Solution 1.* [rt]. Employers are bound by OSHA to provide each employee with a working environment free of recognized hazards that cause or have the potential to cause physical harm or death. Employers must have proper instrumentation for the evaluation of test data provided by an expert in the area of industrial hygiene. This instrumentation must be obtained because the presence of health hazards cannot be evaluated by visual inspection. This data collection effort provides the employer with substantial evidence to disprove invalid complaints lodged by employees alleging a hazardous working situation. This law also gives employers the right to take full disciplinary action against those employees who violate safe working practices in the work place.

2. *Health and Hazard Risk Assessment Solution 2.* [rt]. NIOSH recommends standards for industrial exposure that OSHA uses in its regulations. OSHA has the power to enforce all safety and health regulations and standards recommended by NIOSH.

3. *Health and Hazard Risk Assessment Solution 3.* [pcy].

    1. Hazard Identification. The determination of whether or not a particular chemical can cause an adverse effect. It can define the hazard and nature of the harm.
    2. Dose-Response Assessment. The determination of the probability of an undesirable health response to a given chemical dose.
    3. Exposure Assessment. The determination of the concentration of a contaminating agent in the environment and the estimation of its rate of intake in target organisms.
    4. Risk Characterization. Based on information from the previous steps this step provides an estimate of the magnitude of the health problem associated with an actual exposure event.

4. *Health and Hazard Risk Assessment Solution 4.* [pcy].

    a. Risk Management. After the risk assessment procedures are completed, various options are evaluated in a proposed solution to provide reduction of risk to the exposed population. Specific actions which are identified and selected may include consideration of engineering constraints as well as regulatory, social, political and economic issues related to the exposure.

4. *Health and Hazard Risk Assessment Solution 4.* [pcy] (continued)

    b. Risk Communication. This step is a part of the risk management process that includes exchanging risk information among individuals, groups, and government agencies. The major challenge in this phase of the risk management process is passing risk information from the experienced expert to the non-experienced, but greatly concerned public.

    c. Risk Estimation. This embodies the risk assessment process designed to estimate risk based on:

        1. The nature and extent of the source;
        2. The chain of events, pathways, and processes that connect the cause to the effects; and
        3. The relationship between the characteristics of the impact (dose) and the type of effects (response).

    d. Risk Perception. This term describes an individual's intuitive judgment about risk which is often not in agreement with the level of risk as judged by experts.

    e. Comparative Risk Assessment. The comparison of potential risks associated with a variety of activities and situations so that a specific action can be placed in perspective with other risks. An attempt is often made, for example, to compare an individual's risk of death or cancer from exposure to a hazardous waste site with that associated with traveling in an automobile or eating a peanut butter sandwich (both of which have relatively high risks, but which are perceived very differently from lower risks of hazardous waste sites by the public).

5. *Health and Hazard Risk Assessment Solution 5.* [dls]. Human health risk assessment provides information to bridge the gap between environmental data and risk characterization. Health risk assessments bring order and consistency to the process of dealing with scientific issues relating to the magnitude of a hazard and whether a true hazard exists. They help to answer questions regarding the cost effectiveness of reducing environmental concentrations of toxins and/or reducing exposures to them.

6. *Health and Hazard Risk Assessment Solution 6.* [dls]. Reference dose (RfD) is "an estimate (with an uncertainty of one order of magnitude or more) of a lifetime dose which is likely to be without significant risk to human populations." RfD is determined by dividing the no observed adverse effect level (NOAEL) dose of a substance by the product of the uncertainty and modifying factors as is illustrated in the following equation and in Figure 2 below:

$$RfD = \frac{NOAEL}{(UF)\,(MF)}$$

where UF = uncertainty factor which is generally in multiples of 10 to account for variation in the exposed population (to protect sensitive sub-populations), uncertainties in extrapolating from animals to humans, uncertainties in using sub-chronic instead of chronic study data, and uncertainties in using the lowest observable adverse effect level (LOAEL) rather than the NOAEL; and MF = modifying factor which ranges from 0 to 10 to reflect qualitative professional judgment of additional uncertainties in the data from available studies.

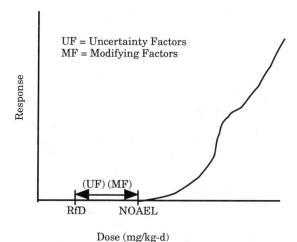

Figure 2. Example dose-response curve showing RfD and NOAEL determinations.

7. *Health and Hazard Risk Assessment Solution 7.* [dls]. For the purposes of an exposure assessment done as part of a risk assessment, EXPOSURE is defined as the contact of a chemical, physical, or biological agent with the OUTER boundary of an organism. It is quantified as the CONCENTRATION of the agent in the medium contacted, integrated over the DURATION of the contact.

8. *Health and Hazard Risk Assessment Solution 8.* [sxl]. The standard values for an "average" person are:

   a.   a body weight of 70 kg for an adult or 10 kg for a child

   b.   drinks 2 L per day of water for an adult or 1 L per day for a child

   c.   breathes 20 m^3 of air per day for an adult or 10 m^3 per day for a child

   d.   a dermal contact area of 1000 cm^2 for an adult or 300 cm^2 for a child

   e.   a life span of 70 years for an adult

9. *Health and Hazard Risk Assessment Solution 9.* [dls]. The averaging time (AT) typically used for acute effects is 1 day. For carcinogenic effects, an exposure duration of 70 years is typically used as the AT.

10. *Health and Hazard Risk Assessment Solution 10.* [dls]. Mathematical predictive computer modeling should be used when there is a lack of adequate monitoring data at the exposure point, when an improved understanding of the relative importance of various routes of exposure is needed, when an improved understanding of the exposure setting is needed, and/or when Monte Carlo analysis of uncertainty in the exposure estimates is desired.

11. *Health and Hazard Risk Assessment Solution 11.* [dls]. Uncertainty may be defined as the lack of knowledge about the correctness of an exposure assessment. Exposure uncertainties arise from a lack of knowledge about the concentrations of the substance(s) of interest in the media that the population is exposed to, and in the duration of the exposure to the substance(s). Examples of information which contribute to uncertainty include wind direction and wind speed which affect dispersion and inhalation exposure in the lower atmosphere, source of drinking water and the amount of water actually consumed for ingestion exposure estimates, and the amount of contaminant(s) in vegetables and the amount of contaminated vegetable matter consumed for ingestion exposure estimates.

12. *Health and Hazard Risk Assessment Solution 12.* [wma]. Uncertainties associated with determining the "Human Exposure" term include the following:

     •   Limited knowledge of source characteristics, e.g., how much of a contaminant is released and for how long has the release occurred.

12. *Health and Hazard Risk Assessment Solution 12.* [wma] (continued)

- Difficulty in describing and calculating the "fate and transport" of the contaminant as it travels from release point in the environment to the exposed population, i.e., the receptor (sometimes referred to as pathway analysis).
- Mobility of the exposed individual or population, thereby constantly changing the individual's exposure.

Uncertainties associated with determining the "Potency of Contaminant" term include the following:

- Potency factors are often based on animal toxicity studies, with uncertain applicability to humans.
- Variable effects on exposed humans due to differences in age, sex, health condition, etc.
- Extrapolation from measured high dose effects to calculated low dose effects.

13. *Health and Hazard Risk Assessment Solution 13.* [dls]. U.S. EPA's Risk Assessment Guidance for Superfund (U.S. EPA, 1989) lists the following questions that should be answered in the evaluation of the fate and transport of contaminants.

- What are the principal mechanisms for change or removal in each of the environmental media?
- How does the chemical behave in air, water, soil, and biological media? Does it bioaccumulate or biodegrade? Is it absorbed or taken up by plants?
- Does the agent react with other compounds in the environment?
- Is there intermedia transfer? What are the mechanisms for intermedia transfer? What are the rates of intermedia transfer or reaction mechanisms?
- How long might the chemical remain in each environmental medium? How does its concentration change with time in each medium?
- What are the products into which the agent might degrade or change in the environment? Are these products potentially of concern?
- Is a steady-state concentration distribution achieved in the environment or the specific segments of the environment?

13. *Health and Hazard Risk Assessment Solution 13.* [dls] (continued)

U.S. EPA, *Risk Assessment Guidance for Superfund. Volume 1. Human Health Evaluation Manual (Part A),* U.S. EPA, Washington, D.C., 1989, PB90-155581,National Technical Information Service, Springfield, Virginia.

14. *Health and Hazard Risk Assessment Solution 14.* [dls]. Assuming that those exposed can be represented by a 70 kg individual, then the volume of drinking water at the MCL to reach the RfD for atrazine is:

$$(3.5 \text{ mg/kg/d}) (70 \text{ kg})/(0.003 \text{ mg/L}) = 81,667 \text{ L/d}$$

This large volume indicates that there is considerable uncertainty (i.e., the product of the uncertainty factors is large) in estimating a reference dose for atrazine.

15. *Health and Hazard Risk Assessment Solution 15.* [dls]. The following figure shows the completed boxes and all of the four basic components comprising a human health risk assessment.

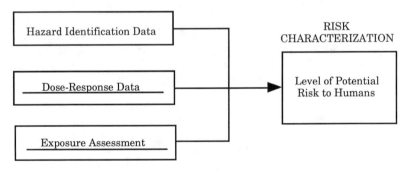

Figure 3. Relationship among basic elements of a human health risk assessment, completed figure.

16. *Health and Hazard Risk Assessment Solution 16.* [dls]. Monte Carlo simulation is a mathematical approach which uses random numbers to describe the effects of uncertainty in a computer model. Given a range of possible values for variables used in the risk assessment process, and the distribution of the frequency with which those values might occur, Monte Carlo simulation can produce a risk forecast chart showing the range of possible risks and the likelihood of their occurrence.

17. *Health and Hazard Risk Assessment Solution 17.* [dls]. The carcinogenic "unit risk" is the risk per unit concentration of the substance in the medium where human contact occurs.

18. *Health and Hazard Risk Assessment Solution 18.* [dls].
Using the ingestion rate equation given in the problem statement,
the following benzene intake rate is determined

$$\text{Ingestion Rate (mg/kg-d)} = \frac{(1 \text{ mg/L}) (2 \text{ L/d}) (350 \text{ d/yr}) (30 \text{ yr})}{(70 \text{ kg}) (70 \text{ yr})}$$

$$\text{Ingestion Rate} = 0.012 \text{ mg/kg-d}$$

The cancer risk from this ingestion exposure at half of the odor
threshold of 1 mg/L is calculated based on the benzene unit risk of
$8.3 \times 10^{-7}/(\mu g/L)$ or:

$$\text{Cancer risk} = (1{,}000 \text{ } \mu g/L) [8.3 \times 10^{-7}/(\mu g/L)] = 8.3 \times 10^{-4}$$

This risk is high relative to the widely accepted standard range
of environmental risk of 1×10^{-6} to 1×10^{-4}.

19. *Health and Hazard Risk Assessment Solution 19.* [dj]. The
following are some of the factors that a consulting engineer should
consider in conducting the exposure and risk assessment:

1. Presence or absence of a basement, and if present, possible ingress routes for liquid and vapor into the basement.
2. Ventilation of the dwelling.
3. Concentration of the dissolved hydrocarbon fraction in groundwater near the dwelling.
4. Presence or absence of separate-phase gasoline near the home.
5. Water supply system used by the dwelling's occupants; i.e., are they using a well that taps into the contaminated aquifer, or are they on a municipal water distribution system?
6. Ages, genders, and lifestyles of the dwelling's occupants.
7. Concentration of vapors in the vadose zone near or under the house, and possible risk of explosion or intoxication.
8. If not yet near the dwelling, the rate of transport of dissolved and separate phases towards the dwelling.
9. Toxicity and cancer dose-response data for the constituents of the gasoline.
10. Estimated additional cancer risk for dwelling's occupants when exposure data are combined with cancer dose-response data.

20. *Health and Hazard Risk Assessment Solution 20.* [jr]. Based on the equation given in the problem statement for estimating action levels for carcinogens, the action level for benzene for an 80 kg person with a life expectancy of 70 years and an exposure duration of 15 years is:

$$Cm = \frac{(10^{-6})\,(80\ kg)\,(70\ yr)}{(1.80\ kg\text{-}d/mg)\,(15\ m^3/d)\,(0.75)\,(15\ yr)}$$

$$Cm = 0.0184\ \mu g/m^3$$

If this person is exposed to a concentration of benzene greater than 0.0184 $\mu g/m^3$ over a 15-year period, the risk is not "acceptable" and remedial action should be taken.

21. *Health and Hazard Risk Assessment Solution 21.* [jr]

a. The benzene concentration in the storage room is given by:

$$Ce = (0.05)\,(1\ kg/10\ m^3)\,(1,000,000\ mg/kg) = 5 \times 10^3\ mg/m^3$$

b. The concentration in Part a is considered to be the action level, Cm, of $5 \times 10^3\ mg/m^3$. Substituting this value into the risk assessment equation given in Problem 20 yields:

$$5 \times 10^3\,mg/m^3 = \frac{(0.1)\,(80\ kg)\,(70\ yr)}{(1.8\ kg\text{-}d/mg)\,(15\ m^3/d)\,(0.75)\,(ED)}$$

Solving for the exposure duration, ED, results in the following:

$$ED = (5.53 \times 10^{-4}\ yr)\,(365\ d/yr)\,(24\ hr/d) = 48.5\ hr$$

22. *Health and Hazard Risk Assessment Solution 22.* [jr]

a. From the output report, at a distance of 0.5 km directly downwind of the stack, the predicted concentration of ethyl-glycol-methyl ether is 19.3 $\mu g/m^3$.

b. Since the acute REL for this chemical (370 $\mu g/m^3$) is almost 20 times as much as the actual exposure level (19.3 $\mu g/m^3$), the person is not at risk.

22. *Health and Hazard Risk Assessment Solution 22.* [jr] (continued)

    c. By using the "discrete distances" option of the program, at a distance of about 0.880 km, the concentration is 364.9 µg/m³ (and increasing with distance). At a distance of 5.9 km, the concentration is 369.5 µg/m³ (and decreasing with distance). The "risk range" is therefore about 0.88 to 5.9 km downwind from the stack.

23. *Health and Hazard Risk Assessment Solution 23.* [dj]. Title III of the Superfund Amendments and Reauthorization Act (SARA) is a free-standing piece of legislation that is also called the Emergency Planning and Community Right-to-Know-Act of 1986. This act requires county, state, and federal government agencies to work with industries to develop "community right-to-know" reporting on hazardous chemicals. Industries are required to report to local emergency planning agencies regarding the quantities and locations of hazardous chemicals stored at their facilities. Government agencies and industry are then to develop emergency preparedness plans that are put into action if chemical emergencies occur at these industrial storage and production facilities.

24. *Health and Hazard Risk Assessment Solution 24.* [dj]. An example of three specific types of chemical emergencies that could occur in a large metropolitan area could include:

    1. Release of a cloud of gas from the spill of 1,000 kg of hazardous chemical from a railroad tanker car.
    2. Explosion and fire at a petrochemical plant.
    3. Leakage of a corrosive acid from storage drums into a nearby river or lake.

This list is provided as an example of the types of industrial accidents that are common in most major, and even smaller, industrialized American cities. Students are encouraged to review local and national newspapers for current examples of such chemical accidents and emergencies.

25. *Health and Hazard Risk Assessment Solution 25.* [dj]. Five typical features of an emergency preparedness plan include:

    1. Evacuation plans and routes for nearby dwellings, schools, offices, and industries.
    2. Access routes and predicted response times for emergency teams.

25. *Health and Hazard Risk Assessment Solution 25.* [dj] (continued)

3. Ability to predict the trajectory and concentration of airborne or waterborne toxic releases.
4. Ability to provide emergency services to the site; for example, adequate municipal water flow rate and pressure to extinguish a major petrochemical fire.
5. Plans for mobilization of area medical personnel to treat casualties associated with major airborne accidental chemical releases.

Additional components of emergency management plans include:

- Methods for communication of risks and evacuation information, etc., to the general public during an emergency.
- Identification of roles and responsibilities of participating entities in responding to and controlling chemical hazards.
- Mechanisms for the testing and updating of emergency management plans through mock drills and exercises.
- Mechanisms for training of professional and voluntary response personnel to make sure the plan and services provided by response personnel are continually updated.
- Methods for the review of performance of personnel involved in real and/or "practice" emergencies so planning and response procedures can be updated and improved over time.

Students are encouraged to discuss additional requirements of effective emergency management plans so that they may identify additional attributes of these plans that have not been listed above.

Chapter 3

RISK COMMUNICATION

Bill Kroesser, David James, and Scott Lowe

I. INTRODUCTION

The last three decades have shown increased public awareness of the threats posed by environmental problems and heightened expectations that the public will be protected from harm. Further, it is now expected that the public will be involved in decisions affecting their safety. As the U.S. EPA supports this concept, environmental engineers and scientists must be ready to effectively communicate with the public regarding environmental risks.

This chapter touches on the major issues regarding communication of environmental risks, and presents several scenarios in which risk communication is important. Risk itself is a difficult concept to understand, even for the engineer or scientist. Can one fully comprehend the fact that each of us has a 0.000035 chance of being hit by lightning during our lives, or that one would need to buy a lottery ticket once a year to have a 0.000035 chance of winning a million dollars?

For the public, such concepts are even more difficult to understand. Thus, engineers and scientists may be tempted to avoid involving these non-experts in the process. Even when persuaded that the public must be involved, how does one explain the hazards in non-technical terms? Once we properly take into account the likelihood of an accident, and sum up the consequences for all who would be affected, how would one take into account the public's fear and outrage this accident might cause? And once we have understood that these are valid components of a risk management plan, how do we carry out a community outreach program to build trust within the community, and obtain the community's feedback?

While the answers to these questions are not easy to develop, and risk communication can be less of an exact science than other aspects of the risk management process, there is a high likelihood that an important project will fail or be considerably delayed if an effective process for the communication and discussion of risk is not competently implemented.

The problems in this chapter cover risk communication related primarily to health risk assessments, emphasizing community outreach, public notification, comparison of relative risks, and analysis of the sometimes large discrepancies between risk perception by the public and that by technical experts.

43

II. PROBLEMS

1. *Risk Communication Problem 1.* (cardinal rules, risk communication). [cmd]. While there is no easy prescription for successful environmental risk communication, practitioners in the field generally agree that there are seven "cardinal rules" of risk communication. List these seven cardinal rules along with the guidelines that risk communication practitioners should follow when communicating risk to the concerned public.

2. *Risk Communication Problem 2.* (environmental risk). [cmd]. What three pieces of information does the public need in order to make informed decisions about environmental risk and risk reduction?

3. *Risk Communication Problem 3.* (community outreach program). [cmd]. Identify the elements of a comprehensive community outreach program for environmental risk communication.

4. *Risk Communication Problem 4.* (environmental risk, environmental hazard). [cmd]. What is meant by "environmental risk"?

5. *Risk Communication Problem 5.* (Acceptable risk, risk communication, reliability). [sn]. As a well educated person in hazardous waste management attending a public hearing on a risk assessment study where the outcome is "no potential health hazard to the public," what are the main concerns you would have in order to be convinced that the results are reliable and acceptable?

6. *Risk Communication Problem 6.* (toxic chemicals, emissions, exposure). [cmd]. If you were a resident of Ourcity, what issues or questions would you be concerned about after reading the following newspaper article?

> 325,000 Pounds of Four Toxic Chemicals Emitted
> Locally - Benzene, Chlorine, Pyridine, Ammonia
> Most Prominent. Industry Says, "Risk is Low."

Last year, 15 local manufacturing facilities emitted more than 10,000 tons of toxic chemicals into the air, water, and land of Ourcity. The top chemicals emitted (in pounds) were benzene (200,000), chlorine (100,000), pyridine (10,000), and ammonia (15,000).

Benzene is a known carcinogen. Chlorine is a highly toxic chemical that may cause severe respiratory problems. Chlorine was involved in the recent accident at North High School, causing evacuation of 1100 students and teachers.

6. *Risk Communication Problem 6*. [cmd] (continued)

Pyridine is a reproductive toxin, causing possible damage to reproductive organs, as well as having serious effects on the central nervous system.

Ammonia, a common household cleaner, is irritating to eyes and the respiratory system.

Newspaper staff examined reports submitted by 15 local manufacturing plants under the requirements of a federal law, the Emergency Planning and Community Right-to-Know Act. The U.S. EPA requires plants to disclose the amount of toxic chemicals they release into the environment each year.

In addition to benzene, chlorine, pyridine, and ammonia, local facilities emit more than 500,000 pounds per year of ethylene, creosols, formaldehyde, and 12 other chemicals.

Tom Jones, senior safety engineer for Newtown Chemical Company, noted that the emissions reported do not give cause for any alarm. Benzene emissions by all 15 companies, he said, are only one-tenth of the benzene given off by automobiles in Ourcity. Jones also pointed to a recent study by the State Environmental Department which showed that total concentrations of benzene and seven other chemicals in Ourcity are well below state standards. In Ourcity, they have been measured at about 20 parts per billion at the intersections of Broad and Main Streets.

Rodney Smith of the State Environmental Department stated that the department will be looking more closely at the emissions to see whether they violate any state standards. "For now," he said, "we are just happy to see the companies providing the reports and complying with the law. Later we will use the data to examine whether we need regulatory changes."

7. *Risk Communication Problem 7*. (expert opinion, outrage factors). [cmd]. Environmental risks are often perceived very differently by the affected population and the "experts" who must characterize the risk and propose management solutions to lessen these risks to the environment. Experts, because they do not have to live with these environmental risks, often communicate the risk in terms of "facts and figures." To the affected population, these data often evoke responses collectively called "outrage factors." Identify and discuss some of these factors.

Reference: Holmes, G., Singh, B. and Theodore, L., Environmental risk assessment, in *Handbook of Environmental Management and Technology*, Wiley-Interscience, New York, 1993, pp. 584-585.

8. *Risk Communication Problem 8.* (environmental risk, home, workplace). [cmd]. Should the general public know about the environmental risks associated with their places of work and/or residence?

9. *Risk Communication Problem 9.* (risk comparisons). [cmd]. One frequently used method of risk communication is that of comparing the risks at a site or with a process or activity to those associated with other common sources of risk, such as driving an automobile or smoking. What are the problems associated with this approach to discussing risk?

Reference: **LaGoy, P. K.**, *Risk Assessment*, Noyes Publications, Park Ridge, New Jersey, 1994, pp. 120-121.

10. *Risk Communication Problem 10.* (ranking of risk, experts). [cmd]. Does the general population rank risk the same way the "experts" do?

Reference: **LaGoy, P. K.**, *Risk Assessment*, Noyes Publications, Park Ridge, New Jersey, 1994, pp. 116-117.

11. *Risk Communication Problem 11.* (greenhouse effect). [mh]. The issue of global warming (the greenhouse effect) has deeply divided the environmental community. Compare the following statements from well-known spokespersons from each side. Has either person helped settle the issue based on these statements?

Dr. Jastrow: "There have been charges that the earth's temperature has followed changes in solar activity over the last 100 years. When solar activity increased from the 1880s to the 1940s, global temperatures increased. When it declined from the 1940s to the 1960s, temperatures also declined. When solar activity and sunspot numbers started to move up again in the 1970s and 1980s, temperatures did the same."

Dr. James Hansen: "Firstly, I believe the earth is getting warmer, and I can say that with 99% confidence. Secondly, with a high degree of confidence, we can associate the warming and the greenhouse effect. Thirdly, in our climate model, by the 1980s and 1990s, there is a noticeable increase in the frequency of drought."

Reference: **Holmes, G., Singh, B. and Theodore, L.**, Environmental risk assessment, in *Handbook of Environmental Management and Technology*, Wiley-Interscience, New York, 1993, pp. 128-130.

12. *Risk Communication Problem 12*. (worst case scenario, risk communication, air quality). [fwk]. Ammonia is the number one chemical involved in reportable accidental release incidences. At a local manufacturing plant, a cylindrical tank is used to store 1 million pounds of anhydrous ammonia, in liquid form, at 70° F. The neighbors want to know the consequences of a leak in the tank. U.S. EPA has asked that a Risk Management Plan be produced, which describes these consequences quantitatively. While a full-scale experiment could be carried out, say in a desert, you are being asked to instead predict the consequences. (A team approach is recommended.)

The liquid occupies half of the volume of the tank. The tank has a pipe which enters below the liquid level, and another above the liquid level.

    a.   What is the pressure in the tank? What is the boiling point of ammonia? Explain these in terms that the public would understand.

    b.   Suppose that the tank suffered a rupture that tore it in half, releasing all of its contents at once. What would be the temperature of the ammonia immediately after the rupture? What fraction of the ammonia would be in liquid form? Describe the thermodynamic changes that would occur in such a circumstance.

    c.   Describe for the neighbors the other kinds of leaks that might occur, and describe the state of the ammonia as it leaves the tank. For example, a partially opened valve on the upper pipe would cause gaseous ammonia to be emitted at a rate determined by the extent of the opening of the valve, while a clean break in the upper pipe would cause choked flow from the pipe.

    d.   For leaks through the upper pipe, describe the changes in temperature and pressure that would occur in the tank as the leak progressed.

    e.   For leaks through the lower pipe, describe the changes in temperature and pressure that would occur in the tank as the leak progressed.

13. *Risk Communication Problem 13*. (health risk assessment, epidemiology, risk communication). [fwk] You are the president of the Town Council for the town of Collegeville, in the state of West Erie. There is a facility in Collegeville at which waste material is deposited. Hereinafter, this facility will be affectionately referred to as the "dump." The Town Council is responsible for regulating such operations; in this exercise there are no country, state, nor

13. *Risk Communication Problem 13*. [fwk] (continued)

federal governmental agencies or bureaus that currently deal with this kind of facility.

There has been some citizen concern raised in one of Collegeville's neighborhoods, an area just downwind of the dump. A study by a national nonprofit group reports that the incidence of illness in that neighborhood is considerably higher than the U.S. average illness.

The people who live there are worried. They feel trapped. They are mostly elderly people, from working class backgrounds, without the means to buy a house elsewhere. Many are second-generation immigrants who have lived there all of their lives. They want the facility to be closed, and the material there to be removed.

Before you assume that the facility causes these illnesses, what kinds of additional information or studies would you request? Why?

Be specific! Do not suggest something as vague as "further health studies." You might, for example, decide to find out whether the illnesses occur more often in this neighborhood than in a "clean" neighborhood across town. The suspicion here would be that the high rate of illness might exist for all West Erieans, regardless of how far they live from the offending facility.

14. *Risk Communication Problem 14*. (health risk assessment, risk communication, carcinogens). [fwk] The following appeared in the newspaper some time ago:

> "WASHINGTON (AP) - Toxaphene, one of the most widely used agricultural pesticides, causes liver cancer in mice and should be considered at least a potential hazard in humans, the National Cancer Institute said yesterday.
>
> The government is weighing the risks and benefits of the insecticide to see if it should be restricted after 30 years of use.
>
> The sole producer of toxaphene said the test results had little significance for human health because of the high doses used. It said the chemical should be tested at the normal amounts used in field conditions."

Imagine that this article was published very recently, and that there is considerable concern regarding the implications of these very tentative results. You are assigned to work on this problem for the West Erie Protection Agency (WEPA). You have been given access for 6 months to any three of the scientists available at WEPA or another cooperating government agency. You need to have them do the following:

14. *Risk Communication Problem 14*. [fwk] (continued)

- Identify the hazard
- Develop a dose-response relationship
- Assess the exposure of the people involved

What type of tasks would you give them? What uncertainties would exist in their results? Describe in detail what you want them to do. If 6 months is not enough time, include a justification for a request for more time.

15. *Risk Communication Problem 15*. (health risk assessment, Monte Carlo simulation, risk descriptors). [dls]. U.S. EPA Region VIII's Superfund Technical Guidance of September 1995 entitled *"Use of Monte Carlo Simulation in Risk Assessments"* states that "Recent EPA guidance recommends developing 'multiple descriptors' of risk to provide more complete information to Agency decision makers and the public." What is the value of having 'multiple descriptors' of risk in human health risk assessments?

16. *Risk Communication Problem 16*. (health risk assessment, hazard quotient, Hazard Index). [dls].

    a. In a health risk assessment, a hazard quotient may be calculated to assess an individual's exposure to a hazardous compound. Write the general mathematical relationship that is used to calculate a hazard quotient, and explain how a hazard quotient is interpreted.

    b. A Hazard Index is related to the hazard quotient. Give an appropriate definition of the Hazard Index.

17. *Risk Communication Problem 17*. (health risk assessment, dose-response, exposure assessment, risk characterization, acceptable risk). [dls]. What factors may eventually lead to identification of levels of acceptable risk being defined on a case-by-case basis?

18. *Risk Communication Problem 18*. (emergency planning, public notification). [fwk]. The task of achieving effective risk communication is difficult because the public often lacks important and relevant information about the risks posed by local manufacturing and storage facilities. In order to alleviate this problem, various laws have been enacted to increase the amount of information available to regulators, planners, and the public. Discuss the most important of these, one of which was enacted in 1986, and the other, in 1990.

III. SOLUTIONS

1. *Risk Communication Solution 1*. [cmd].   The risk communication practitioner should follow these seven cardinal rules of risk communication:

1. Accept and involve the public as a legitimate partner.
   a. Involve the community early in the process.
   b. Identify and involve community "stakeholders."
   c. Remember that the public may hold you accountable for the outcomes.
2. Plan carefully and evaluate your efforts.
   a. Begin with clear, explicit risk communication objectives.
   b. Evaluate the information you have about the risks and know its strengths and weaknesses.
   c. Aim your communications at specific subgroups in your audience.
   d. Utilize competent spokespersons and staff to assist in communication.
   e. Pretest your messages and learn from your mistakes.
3. Listen to the public's specific concerns.
   a. Utilize a variety of techniques to find out what people are really thinking.
   b. Empathize with your audience and use this to help you understand their concerns.
   c. Identify "hidden agendas" or other political considerations that can complicate the task of communication.
4. Be honest, frank, and open.
   a. State your credentials but do not expect to be trusted immediately, if ever.
   b. If you do not know an answer, admit it, and then get back to people with answers.
   c. Disclose risk information as soon as possible.
   d. Do not minimize or exaggerate the level of risk but be as open as possible.
   e. Discuss strengths and weaknesses of data sources.
5. Coordinate and collaborate with other credible sources.
   a. Take time to coordinate all inter- and intragovernmental communications.

1. *Risk Communication Solution* 1. [cmd] (continued)

  b. Devote effort and resources to building bridges with other organizations.
  c. Consult with others to determine who is best able to answer questions about risk.
  d. Utilize "joint communications" with other trustworthy sources where possible.

 6. Meet the needs of the media.
  a. Be accessible and cooperative with reporters.
  b. Prepare in advance and provide background material and graphics on complex risk issues.
  c. Do not hesitate to follow up on stories with praise or criticism, as warranted.
  d. Work to build a state of mutual trust with reporters and editors.

 7. Speak clearly and with compassion.
  a. Use simple, non-technical language.
  b. Use examples and concrete images to make the risk data come alive.
  c. Acknowledge and respond to emotions and views that people express.
  d. Promise only what you can do and be sure to do what you promise.

2. *Risk Communication Solution 2.* [cmd]. The public needs to know the following items in order to make informed decisions about environmental risk and risk reduction:

 1. How serious is the risk?
 2. What is and can be done about controlling and reducing the risk?
 3. What will it cost to eliminate the risk?

3. *Risk Communication Solution 3.* [cmd]. The elements of a comprehensive community outreach program for risk communication include:

- Information and education.
- Development of a receptive audience.
- Disaster warning and emergency information.
- Mediation and conflict resolution.

4. *Risk Communication Solution 4.* [cmd]. Environmental risk refers to the likelihood of injury, disease, or death resulting from human exposure to a potential environmental hazard.

5. *Risk Communication Solution 5.* [sn]. Risk assessment and conveying its results to inform the public are controversial and hard to accomplish since the public often does not comprehend the interpretation of risk assessment results in the same way as the risk assessors do. Several factors should be considered in determining the level of acceptability of the results of a risk assessment study. First, the level of acceptable risk used as the end point of the study must be agreed upon as the acceptable level of risk depends upon the individual's level of concern. In addition, in considering reliability and acceptability of the study, many elements leading to the conclusion of the study can be subject to criticism including the capability of the principal investigator, the number and the appropriateness of the subjects used in the health studies, the method employed in assessing health effects, the nature of samples collected, and the detection limits of examined parameters.

- Professional experience is an extrinsic indication of the capability of the investigator to accomplish the study while qualification surmises the principal knowledge of the individual.
- The number of subjects being studied is statistically important in a quantitative risk assessment since the result of a risk assessment is usually reported in terms of the probability that the event might take place. It is therefore important to design into the study an adequate number of samples in order to obtain the required level of significance and reduce the chance of drawing a wrong conclusion (such as to incorrectly reject or fail to reject the hypothesis).
- Suitability and characteristics of the subjects include their sex, age, and susceptibility to health effects from the contaminant(s) of concern.
- Since most of the toxic substances that may affect human health are often measured at low concentrations (e.g., parts per billion or parts per trillion), methods of analysis and detection limits of the parameters will be the key decisions for determining the reliability of the data. Quality control and quality assurance also play important roles in quantifying the accuracy and reproducibility of the data.

6. *Risk Communication Solution 6.* [cmd]. This open-ended question will be answered according to the opinions of the respondent. Several items to consider include:

- The true hazard of the chemicals listed and the way they are described in the article. Ammonia is highly toxic, for example, even though it is used in household cleaners.
- The relative nature of the chemicals listed in terms of their effects on humans. Benzene is a carcinogen, for example, while chlorine is a toxic gas and is a strong oxidizer.
- The comparison of the amount of benzene released by the 15 chemical companies compared to all of the automobiles in the city should be evaluated carefully. If the city is large, these 15 emitters could each represent a large number of automobile "equivalents" and may represent a very significant emission source. This indicates as well that controls on these 15 sources may provide a cost effective way of reducing the release of the carcinogen, benzene, into the atmosphere.
- The true release of benzene from these automobiles should be checked. With modifications of fuels to reduce their benzene levels, with pressurized fuel systems, and mandated vapor controls in some states, this estimate of benzene emissions from automobiles may not be accurate.
- Why is the State Environmental Department only happy that these companies are complying with the law? Should they not be doing at least that?
- If the State has had this emission information for some time, why haven't they already made progress on evaluating the need for regulatory change?

7. *Risk Communication Solution 7.* [cmd]. Risk perception scholars have identified more than 20 "outrage factors." Here are a few of the main ones:

1. Voluntariness - A voluntary risk is much more acceptable to people than a coerced risk because it generates no outrage. Consider the difference between getting pushed down a mountain on slippery sticks by someone and deciding to go skiing.

7. *Risk Communication Solution 7.* [cmd] (continued)

2. Control - Almost everybody feels safer driving than "riding shotgun." When prevention and mitigation are in the individual's hands, the risk (though not the hazard) is perceived to be much lower than when they are in the hands of a government agency.

3. Fairness - People who must endure greater risks than their neighbors, without access to greater benefits, are naturally outraged, especially if the rationale for so burdening them looks more like politics than science.

4. Process - Does the agency come across as trustworthy or dishonest, concerned or arrogant? Does it tell the community what is going on before the real decisions are made? Does it listen and respond to community concerns?

5. Morality - American society has decided over the last two decades that pollution isn't just harmful, it's evil. Talking about cost/risk tradeoffs sounds very callous when the risk is morally repugnant. Imagine a police chief insisting that an occasional child-molester is an "acceptable risk."

6. Familiarity - Exotic, high-tech facilities provoke more outrage than do familiar risks (your home, your car, your jar of peanut butter).

7. Memorability - A memorable accident, for example, Love Canal, Bhopal, or Times Beach, makes the risk easier to imagine and thus, as we have defined the term, more risky. A potent symbol, the 55-gallon drum, can do the same thing.

8. Dread - Some illnesses are more dreaded than others; compare AIDS and cancer with, say, emphysema. The long latency of most cancers and the undetectability of most carcinogens add to the dread.

9. Diffusion in Space and Time - For example, Hazard A kills 50 anonymous people a year across the country. Hazard B has one chance in 10 of wiping out its neighborhood of 5,000 people sometime in the next decade. Risk assessment tells us the two have the same expected annual mortality: 50. "Outrage assessment" tells us that A is probably going to be acceptable while B is certainly not.

8. *Risk Communication Solution 8*. [cmd]. This is an open-ended question with a number of potential answers depending on the respondent's personal feelings. The following items might be considered in a typical answer:

- Indoor air pollution is not currently regulated by federal laws in residential or commercial buildings, but many organic compounds in these indoor environments represent significant health hazards to those exposed to them.
- "Sick Building Syndrome" has been documented in many office workers, and manifests itself in dizziness, nausea, headaches, etc.
- Strict rules for worker safety have been established by OSHA in industrial workplace environments to protect these workers from excessive exposure to organic and inorganic chemicals.
- The U.S. has a general policy of "informed consent" suggesting that it is both proper and necessary to divulge information to the general public relating to the environmental risks they may be exposed to at their places of work or residences.

9. *Risk Communication Solution 9*. [cmd]. A frequently used approach to addressing risk concerns is to compare the risks at a site with those associated with other common sources of risk. The problem with this type of comparison is that although objectively similar, i.e., both the site and smoking, for example, pose a risk of developing cancer, they are subjectively quite different. The risks differ in being involuntarily versus voluntarily imposed, being unfamiliar versus familiar, and the individual having no control versus having control over the initiation of the risks. A review of outrage factors discussed in Problem 7 may provide some additional insights into the problem with many risk comparisons that are made.

10. *Risk Communication Solution 10*. [cmd]. Scientists deal in "objective" truths but issues of perception are often highly subjective. Consequently, scientists may present the "correct" answer to a given problem, without understanding that the message perceived by the audience is somewhat (or substantially) different.

As an example, informing an audience at a town meeting that there is no need to be concerned about health risks because soils in their yards only contain dioxins at a level that poses a 10^{-6} cancer risk is likely to result in a different opinion on the part of the audience. The presenter has determined (and probably accurately

10. *Risk Communication Solution 10.* [cmd] (continued)

based on current policy) that this level of dioxin presents a *de minimis* or insignificant risk. If the audience had been a roomful of like-thinking researchers, the response would have probably involved a substantial number of heads nodding in agreement. However, the audience noted above has a somewhat different knowledge base from which to form their risk decisions. The message that is perceived by this audience is that dioxin, the deadly toxic compound and a carcinogen, is present in their backyards. The answer to the question, "Does the presence of this chemical pose a risk?" is therefore substantially different from the answer suggested by the speaker. In addition, the presentation of the message, i.e., "there is no need for you to be concerned," is patronizing and inaccurate in light of the background, knowledge base, and overall concerns of the people receiving the message.

Acceptable versus unacceptable risks (or risks people are concerned about) include factors such as fears posed by uncertainties in the sciences, concerns about lower property values, etc. Scientists need to constantly remember that the subjective perception of an issue can be quite different than the objective answer.

11. *Risk Communication Solution 11.* [mh]. Those who believe the greenhouse effect is not true are led by the George C. Marshall Institute. These scientists conclude the following:

1. Any warming is best explained by variations in the natural climate and solar energy concentrations of carbon dioxide in the atmosphere.
2. Natural variations in solar activity occur and have an impact on weather and climate variability.
3. The constant tilt of the earth varies from 22° to 24.5° every 41,000 years.
4. German and American scientists have found an 11 year sun spot cycle linked to weather patterns on a global scale.
5. Between 1880 and 1930, scientists observed a 1°F rise in temperature, before 67% of global CO_2 emissions even occurred.
6. Under such cycles, a 2°F decrease is expected to offset a so-called greenhouse effect.

Those who believe the greenhouse effect is real are led by NASA and the U.S. EPA. These scientists and regulators conclude the following:

11. *Risk Communication Solution 11.* [mh] (continued)

    1.   Climate models have been reliable enough to conclude the greenhouse effect is causing global warming.

    2.   An increase of atmospheric CO_2 from 280 ppm to 350 ppm has been observed over the duration of the CO_2 record.

    3.   Dr. Hansen feels most scientists believe in the greenhouse effect but they will not say so because of a lack of conclusive evidence.

    4.   Droughts will develop unless large volcanic eruptions occur which serve to cool the earth.

As one can tell from these diverse statements, there is not a consensus in the scientific community on the greenhouse effect. Each side claims to be correct on this controversial issue. Yet, each statement shows that many environmental policies will not be left up to pure science. The ability to convince others of your findings may be as important as the actual scientific findings themselves. This stresses the fact that communication is essential in solving a future real or perceived problem. It also suggests the need for careful explanation of the rationale for various sides of a scientific debate to reassure the public that decisions can be made based on validated, reliable scientific evidence, and that debate is a healthy and necessary part of the scientific method.

12. *Risk Communication Solution 12.* [fwk].

    a.   Pressure in the tank is the pressure at which saturated ammonia exists at 70°F. From Perry's Chemical Engineering Handbook table on Properties of Saturated Ammonia, p = 128.8 psia at 70°F. From this same table, at p = 14.7 psia, T = -28.2°F.

    b.   A sudden failure of the entire vessel would cause the entire contents to drop from 128.8 psia to 14.7 psia. The reduction in pressure would cause a temperature drop, and the flashing would cause further temperature loss. This would continue until the remaining liquid reached the normal boiling point. The result would be that 82% of the contents would lie in a pool of liquid at its normal boiling point of -28.2 °F, and 18% would be in the form of a cold gas. The liquid would sit in a pool on the ground, and evaporate as heat from the air above and the

12. *Risk Communication Solution 12.* [fwk].

earth below was transferred to the liquid. The boiling of all of the liquid might take hours to be completed. To generate these results, look up the enthalpies of ammonia under the conditions mentioned.

Temp.°F	Enthalpies in BTU/lb	
	liquid	vapor
-28.2	14	602
70	120.5	629

It can be assumed that the change in state occurs in an adiabatic manner. Thus, the enthalpy of the ammonia before the rupture is equal to the enthalpy after the rupture. Before the rupture, the enthalpy is that of liquid at 70°F. After the rupture, the mass fraction of ammonia which vaporizes is x, and the mass fraction that remains as liquid is (1 - x). An enthalpy balance yields:

$$\text{enthalpy before rupture} = \text{enthalpy after rupture}$$
$$120.5 = 14\,(1 - x) + 602\,x$$
$$x = 0.18$$

c. A leak at the end of the upper pipe would result in a gas release to the atmosphere. If there is a clean break in the pipe, the flow rate is governed by the equation for choked flow. As the tank loses ammonia, the pressure would drop until it falls below that needed for choked flow. Then the flow through the pipe would be governed by compressible gas flow equations. The compressible gas flow would also apply if the pipe had enough friction to prevent choked flow.

A leak at the end of the lower pipe would result in a flow of liquid that flashes on entering the atmosphere. The flow rate through the pipe would be governed by laminar or turbulent flow equations for incompressible fluids. In some cases, the flashing may occur inside the pipe. This would result in two-phase flow, and under

12. *Risk Communication Solution 12.* [fwk].

two-phase flow conditions, prediction of the flow rate from the tank pressure and pipe size would be very difficult.

A hole in the upper part of the tank would cause gaseous loss, with flow rates which are predicted by the orifice equation. The pressure drop would cause considerable cooling, possibly resulting in some condensation. A hole in the lower part of the tank would result in liquid flow, following the orifice equation, and flashing of the liquid as it leaves the tank.

d. When a leak occurs in the upper pipe or upper portion of the tank shell, the liquid will boil to replace the lost gas. Eventually, all of the liquid would be evaporated, and the remaining gas would lose pressure as the gas quantity was depleted. However, another important factor should not be overlooked. As the liquid boils, the required heat would most likely come through a lowering of the temperature of the tank contents. This would lower the saturation pressure, and therefore lower the driving force for expulsion of the gas at the leak.

e. When a leak occurs in the lower pipe or lower portion of the tank shell, the loss of liquid will lower the pressure of the vapor above it. As the equilibrium pressure is determined by the saturation conditions, some liquid must boil to counteract the lost pressure. Boiling lowers the temperature of the contents, as described above. Eventually, all of the liquid would be lost, and the emission from the tank would then be in the form of a gas.

13. *Risk Communication Solution 13.* [fwk]. There are many issues raised here. It is difficult to know where we will find useful information. In general, however, there are five categories of questions to raise which include:

1. Were the appropriate variables included in the study? (Age, prior employment exposure, ethnic background, etc.)

13. *Risk Communication Solution 13.* [fwk].

>   2. Were the authors unbiased, or do they have something to gain from acceptance of their results?
>   3. What is the quality of the air, water, food, etc. to which these residents are exposed?
>   4. What materials are stored at the dump? Are any of these known to cause these problems?
>   5. What other sources of risk exist? (Local industries, construction materials of residences)
>   6. Do any of the illnesses traditionally have environmental causes?

A good solution would address as many of these as possible, with sensible recommendations as to what to do. For example, under the category appropriate variables one might wonder if the group studied were older than the comparison group, and whether that could account for the differences seen. If such factors as age, smoking, exposure during employment, etc., were not accounted for properly, a new study might be recommended, or one could carry out a new statistical analysis on the data from this study.

Consider also former employment. A hazardous industry in the neighborhood that closed recently could be the cause of results seen here. Another difficult variable is the ethnic background of those studied. If this is an ethnic neighborhood, is this particular disease one that is seen more commonly in those with this ancestry?

Regarding bias, one might look to see if the authors have an ax to grind.

Sampling of the air, well water, and soil might be suggested as appropriate at this point. It is important to recognize, however, that at this point sampling will not likely solve the problem. There are too many possible offending compounds, and hundreds of them might be found in the air or water. After an expensive exercise in environmental sampling and analysis, toxicologists might decide that no conclusions can be drawn.

Looking at the dump is another possibility, and should be explored, but again without the expectation that this alone will solve the problem. If there is a clear answer at the dump, this might be uncovered by inspecting the facility, looking at a list of what is going into it, or looking at the health of the dump workers. If there is a contaminant at the dump, the workers should be exposed at a higher concentration, and this would likely cause symptoms in these workers at a higher rate than the general population.

13. *Risk Communication Solution 13.* [fwk].

Other sources of risk would include industrial emissions up- and downwind, a locally used food or other product that is contaminated, and household radon, asbestos, formaldehyde, lead, indoor pesticides, and other hazardous household products.

Studying the illnesses is another approach that can yield results. Keep in mind, however, that until we identify the pollutant, epidemiology studies are difficult, and toxicology studies cannot be done at all.

This problem statement was framed to be purposefully vague on the nature of the illnesses in order to focus on the issues outlined above. It could be modified to specify a particular set of symptoms, such as chronic fatigue or headaches, or a verifiable disease, such as mesothelioma. Information regarding any of these diseases can be found on the INTERNET, and likely sources and routes could be selected from this information and the general discussion provided above.

14. *Risk Communication Solution 14.* [fwk]. Six months is clearly not enough time to complete this health risk assessment. A dose-response relationship would normally be done with mice, over their 2-year lifetime, with another 6 months needed to analyze the results.

Assessing the exposure may be difficult because it is not known whether the threat is to manufacturing facility workers, farmers, or those who eat the contaminated food. Bioaccumulation studies might take several years, or more.

The steps required by the government for carrying out a risk assessment are given in many sources. One of the original sources is the booklet, *Risk Assessment and Management: Framework for Decision Making*, U.S. EPA, EPA 600/9-85-002, December, 1984.

15. *Risk Communication Solution 15.* [dls]. Balancing residual risk acceptance by regulators and the affected public against the costs of risk reduction may be done more satisfactorily if the uncertainty of the risk assessment is quantitatively communicated. This kind of information may enhance the sense of control, fairness, and insight to the process (help minimize the so-called "outrage factors") for potentially exposed individuals in communities where the risk exists.

16. *Risk Communication Solution 16.* [dls].

    a. The hazard quotient is defined as the actual dose received by a person divided by the

16. *Risk Communication Solution 16.* [dls].

> reference dose (RfD). A hazard quotient of less than 1 is defined as non-hazardous.
>
> b. A Hazard Index is computed if it is suspected or known that an individual has been exposed to several hazardous compounds. The Hazard Index is defined as the summation of those hazard quotients that are ≥ 1. Hazard quotients with values less than one are not included in the Hazard Index calculation.

17. *Risk Communication Solution 17.* [dls]. Exposed population size, costs of meeting risk targets, and the scientific quality of the risk assessment may all be used to determine the acceptable risk at Superfund sites and other similar contaminated sites.

> **Reference: Graham, J. D.**, The legacy of one in a million, *Risk in Perspective*, 1(1),1-2, 1993.

18. *Risk Communication Solution 18.* [fwk]. In 1986, the Emergency Planning and Community Right-to-Know Act (EPCRA) became Title III of the Superfund Amendments and Reauthorization Act (SARA). As a result, facilities having hazardous materials on site are required to provide to the public a great deal of information regarding the identities and quantities of hazardous materials stored at their facilities. Further, community-led public emergency response programs were to be established.

The Clean Air Act Amendments of 1990 specify that, by 1998, facilities having significant amounts of hazardous materials will need to have completed a Worst Case Scenario analysis, describing the off-site consequences of an accidental release of hazardous material. The intent of the U.S. Congress in passing this law was to stimulate a dialogue between industry and the public to improve accident prevention and emergency response practices.

As a result of these changes, as well as other changes in various laws and regulations, a considerable amount of information is becoming available to the public. This will allow the public to better understand the issues and risks related to the storage and use of hazardous materials in their community. At the same time, the public is becoming more interested in risk issues, and many companies are providing additional information, over and above that required by the law. As this occurs, companies are realizing that public opposition is more closely tied to lack of information than to blind fear and obstinance. They are finding that honesty is paying off, and that the public is willing to accept some level of risk,

18. *Risk Communication Solution 18.* [fwk].

so long as there are no fears that a company is hiding some greater risk from them.

Chapter 4

POLLUTION PREVENTION

Joseph Reynolds

I. INTRODUCTION

The amount of waste generated in the United States has reached staggering proportions. According to the U.S. EPA, 250 million tons of solid waste alone are generated annually. Although both the Resource Conservation and Recovery Act (RCRA) and the Hazardous and Solid Waste Act (HSWA) encourage businesses to minimize the wastes they generate, the majority of current environmental protection efforts are centered around treatment and pollution cleanup, not prevention.

The passage of the Pollution Prevention Act of 1990 has redirected industry's approach to environmental management; pollution prevention (P^2) has now become the environmental option of this decade and the 21st century. Whereas typical waste management strategies concentrate on "end-of-pipe" pollution control, P^2 attempts to handle waste at the source (i.e., source reduction). As waste handling and disposal costs increase, the application of P^2 measures is becoming more attractive than ever before. Industry is currently exploring the advantages of multi-media waste reduction and developing agenda to strengthen environmental design while lessening production costs.

In 1990, Congress established P^2 as a national policy with the Pollution Prevention Act. It is the responsibility of the EPA to enforce that legislation. The Act instituted the following hierarchy of waste management:

1. Source reduction - Practices that decrease, avoid, or eliminate the generation of waste are considered source reduction and can include the implementation of procedures as simple and economical as good housekeeping.
2. Recycling/reuse - Recycling is the use, reuse, or reclamation of wastes and/or materials and may involve the incorporation of waste recovery techniques (e.g., distillation, filtration).
3. Treatment - Treatment involves the destruction or detoxification of wastes into nontoxic or less toxic materials by chemical, biological or physical methods, or any combination of these methods.

4. Ultimate disposal - Disposal has been included in the hierarchy because it is recognized that residual wastes will exist. The EPA's so-called "ultimate disposal" options include landfilling, land farming, ocean dumping, and deep-well injection. However, the term *ultimate disposal* is a misnomer, but is included here because of its adoption by the EPA.

The table below provides a rough timetable demonstrating the national approach to waste management. Note how waste management has begun to shift from pollution control to pollution prevention.

Table 1
Waste Management Timetable

Timeframe	Control
Prior to 1945	No Control
1945–1960	Little Control
1960–1970	Some Control
1970–1975	Greater Control (EPA Founded)
1975–1980	More Sophisticated Control
1980–1985	Beginning of Waste Reduction Management (WRM)
1985–1990	Some WRM Activities
1990–1995	More WRM Activities (Pollution Prevention Act)
1995–2000	WRM Accepted Policy for Industry and Government
2000–	???

In order to properly design and then implement a P^2 program, sources of all wastes must be fully understood and evaluated. A multimedia analysis involves a multifaceted approach. It must not only consider one waste stream but all potentially contaminated media (e.g., air, water, land). Past waste management practices have been concerned primarily with treatment. All too often, such methods solve one waste problem by transferring a contaminant from one medium to another (e.g., air stripping). Such waste shifting is not P^2 or waste reduction.

The previously mentioned multimedia approach to evaluating a product's waste stream(s) aims to ensure that the treatment of a waste stream does not result in the generation or increase in an additional waste output. Clearly, impacts resulting during the production of a product must be evaluated over its entire history or lifecycle. A lifecycle analysis, or "Total Systems Approach," is crucial to identifying opportunities for manufacturing process performance improvement. This type of evaluation identifies "energy use, material inputs, and wastes generated during a product's life: from extraction and processing of raw materials, to manufacture and transport of a product to the marketplace, and, finally, to use and dispose of the product (Theodore and McGuinn, 1992).

The main problem with the traditional type of economic analysis of a P^2 program is that it is difficult to quantify some of the not-so-obvious economic merits. Many proposed P^2 projects have failed to receive funding because their proponents were unable to convince management that they were economically viable. There are several considerations that need to be taken into account in any meaningful economic analysis of a P^2 effort. These factors, listed below, are certain to become an integral part of any P^2 analysis in the future.

- Decreased long-term liabilities
- Regulatory compliance requirements
- Regulatory recordkeeping requirements
- Dealings with the U.S. EPA
- Dealings with state and local regulatory bodies
- Elimination or reduction of fines and penalties
- Potential tax benefits
- Improved customer relations
- Stockholder support (corporate image)
- Improved public image
- Reduced technical support requirement
- Potential insurance costs and claims
- Effect on borrowing power
- Improved mental and physical well-being of employees
- Reduced health maintenance costs
- Improved employee morale
- Other process benefits
- Improved worker safety
- Avoidance of rising costs of waste treatment and/or disposal
- Reduced training costs
- Reduced emergency response planning requirements

Many proposed P^2 programs have been discontinued in their early stages because a comprehensive analysis was not performed. Until the effects described above are included, the true merits of a P^2 program may be clouded by incorrect and/or incomplete economic data.

The problems in this chapter stress the profound opportunities for both the individual and industry to prevent the generation of waste. They cover all aspects of P^2 from waste minimization and source control to ultimate disposal, and focus on improving our ability to quantitatively describe the economic benefits to materials and energy use reductions that result from P^2 efforts.

REFERENCES

The reader is referred to the following text/reference books for additional information on P^2.

1. **Holmes, G., Singh, B. and Theodore, L.**, Environmental risk assessment, in *Handbook of Environmental Management and Technology*, Wiley-Interscience, New York, 1993, chap. 33.
2. **Theodore, L. and McGuinn, Y. C.**, *Pollution Prevention*, Van Nostrand Reinhold, New York, 1992.
3. **Theodore, M. K. and Theodore, L.**, *Major Environmental Issues Facing the 21st Century*, Prentice-Hall, Upper Saddle River, New Jersey, 1995.

II. PROBLEMS

1. *Pollution Prevention Problem 1.* (recycling, pollution prevention). [sxl]. When you go shopping, what kinds of things can you do to save the environment?

2. *Pollution Prevention Problem 2.* (pollution prevention, management). [car]. As the environmental scientist of Company XYZ, you have been given the task of putting together a presentation promoting the development of a company P^2 plan. What are some of the benefits you would present to your company's management to promote the development of this P^2 plan?

3. *Pollution Prevention Problem 3.* (waste minimization assessment). [sn]. You are a consulting engineer hired by a company to evaluate waste minimization opportunities in the organization. Outline your approach for conducting a waste minimization assessment.

4. *Pollution Prevention Problem 4.* (on-site recycling, off-site recycling). [jr]. Recycling may take place either before or after the product reaches the consumer. Recycling prior to product purchases involves raw materials and by-products that are treated and reused during the manufacturing of a given end-product, or that are used for some other purpose.
Over the last decade, the use of solvent recovery has increased due to the increased utilization of industrial solvents. In the interest of P^2, companies often must choose between on-site recycling and off-site recovery of the solvents.

    a.   List some advantages of on-site recovery.
    b.   List some disadvantages of on-site recovery.
    c.   List some advantages of off-site recovery.
    d.   List some disadvantages of off-site recovery.

5. *Pollution Prevention Problem 5.* (landfill, waste composition). [sxl]. The national average composition (weight percent) of materials discarded in municipal solid waste (MSW) is as follows:

paper 41%	yard waste 18%
glass 8%	metal 9%
plastic 9%	food waste 8%
other 7%	

An environmentally conscious city on the East Coast recently has decided to do something about the amount of MSW the city has been dumping into its landfill. The first and logical target for waste reduction is the paper waste which accounts for more than 40 wt% of the total waste generated. After a high-energy campaign drive,

5. *Pollution Prevention Problem 5.* [sxl] (continued)

the amount of the paper going into the landfill decreased by 25%. What is the composition (weight percent) of landfill discards now?

6. *Pollution Prevention Problem 6.* (plastics, recycle). [sn]. Many plastic containers today are stamped with symbols as an aid to recycling. Identify the source of plastic listed below and give examples of containers that are usually produced from each type of material.

    a. 1-PET
    b. 2-HDPE
    c. 3-V
    d. 4-LDPE
    e. 5-PP
    f. 6-PS

7. *Pollution Prevention Problem 7.* (recycling, packaging, pollution prevention). [wma]. A key to identifying and applying P^2 principles is to understand the total life-cycle of a product or material. Identify and discuss potential opportunities for P^2 for a "juice pack" of orange juice. (Hint: Both packaging and content are components of this popular product.)

8. *Pollution Prevention Problem 8.* (break-even costs, paper, plastic, waste). [dj]. A university food service sells 32-ounce (0.946 L) soft drinks in waxed paper cups for $1.15 each. They also sell a reusable, insulated plastic 32-ounce drink container for $3.18 with one free refill of soft drink. Additional refills of the reusable plastic container are $0.94 each. How many soft drinks do you have to buy before you break even on the extra cost of the insulated plastic cup?

9. *Pollution Prevention Problem 9.* (costs, paper, plastic, solid waste). [dj]. Refer to Problem 8 above. How much money would you spend to reach the break-even point?

10. *Pollution Prevention Problem 10.* (costs, paper, plastic, solid waste). [dj]. The expected life of the plastic cup is 1,000 uses. How much money would you save over the life of the plastic cup in terms of costs for drinks from paper cups? Refer to Problem 8 for specific costs for drinks from the paper versus reusable plastic cup.

11. *Pollution Prevention Problem 11.* (costs, paper, plastic, solid waste). [dj]. Refer to Problems 8 and 9. Assume that each waxed paper cup has a mass of 10 g, consisting of 8.0 g of paper and 2.0 g of wax. Each paper cup is crushed to a volume of 50 cm^3 before disposal in a landfill. How much mass (kg) and crushed

11. *Pollution Prevention Problem 11.* [dj] (continued)

volume (m³) of paper cups will have been kept out of the landfill over the 1,000-use life of a single, reusable plastic cup?

12. *Pollution Prevention Problem 12.* (costs, paper, plastic, solid waste). [dj]. Refer to Problems 9 and 11. Assume that an estimated 400,000 joules of energy input during the life of one paper cup is required for its manufacture from bulk paper, transport, use and disposal. There are 142,000,000 joules in a gallon of oil. Based on the results of Problem 8, how much oil will have been used to make the number of waxed paper cups at the reusable plastic/disposable waxed paper cost break-even point?

13. *Pollution Prevention Problem 13.* (costs, paper, plastic, solid waste). [dj]. Refer to Problems 8, 9, 11, and 12. Assume that the plastic cup has a mass of 40 g, consisting of 30 g of polypropylene and 10 g of urethane foam insulation. Assume an estimated energy input of 800,000 joules for its manufacture from plastic pellets, for transport and for selling of the plastic cup, plus an energy cost of 394,000 joules each time the plastic cup is washed with soap and 2 L of warm (60°C) water. At the reusable plastic/disposable waxed paper cost break-even point calculated in Problem 8, how many joules of energy will have been expended to make, transport, use and wash the plastic cup? Does this represent a net energy saving compared to the paper cups?

14. *Pollution Prevention Problem 14.* (costs, paper, plastic, solid waste). [dj]. Refer to Problems 12 and 13. Using methods similar to those shown in the solution to Problem 8, calculate the **energy** break-even point for the plastic versus the wax paper cup.

15. *Pollution Prevention Problem 15.* (costs, paper, plastic, solid waste). [dj]. Refer to Problems 8 through 13. If energy is obtained at an efficiency of 40% from a gallon of oil, calculate how many gallons of oil would be saved over the 1,000-use life of 1,000,000 reusable plastic cups.

16. *Pollution Prevention Problem 16.* (VOCs, emission factors). [jr]. Surface coating entails the deposition of a solid film on a surface through the application of a coating material such as paint, lacquer, or varnish. Surface coating operations are significant emission sources of volatile organic compounds (VOCs). Most coatings contain VOCs which evaporate during the coating application and curing processes, instead of becoming part of the dry film.

A common measure of the level of emission of VOCs is the emission factor, f, which is defined as follows:

16. *Pollution Prevention Problem 16.* [jr] (continued)

$$f = \frac{x_v \, x_w \, \rho_v}{1 - x_s}$$

where f = emission factor, lb_{VOC}/gal_{solids}; x_v = volume fraction of organic volatiles in the solvent; x_w = volume fraction of water in the solvent; x_s = volume fraction of solvent in the coating; and ρ_v = density of organic volatiles, lb/gal.

A canning operation currently uses an organic solvent-based coating which contains 40 vol% organic volatiles. To reduce emissions, it is proposed to replace the organic solvent-based coating with a water-based coating which contains 65 vol% solvent, 80% of which is water. In both coatings, the density of the organic (VOC) portion is 7.36 lb/gal.

    a. Calculate the emission factor for the organic solvent-based coating in lb_{VOC}/gal_{solids}.

    b. Calculate the emission factor for the water solvent-based coating in lb_{VOC}/gal_{solids}.

    c. Determine the percent reduction in VOC emissions that would be achieved if the company switched from the organic to the water solvent-based coating.

17. *Pollution Prevention Problem 17.* [dks]. (material balance, waste segregation, industrial wastewater treatment). Soap, detergents, and various toilet articles are produced in an industry having the sewer map given in Figure 4 below. Fly ash, tank bottoms, and spent caustic are discharged gradually from batch holding tanks into the sewer.

Waste composition data given in Table 2 below are mean values. Using these data, answer the questions below. Sampling stations within the sewer system are denoted in the sewer map (Figure 4).

    a. Carry out a material and flow balance for the plant as it currently exists.

    b. Suggest how wastes could be segregated for treatment. Show your proposal on a revised sewer map.

    c. Prepare a new material and flow balance using the revised sewer layout.

17. *Pollution Prevention Problem 17.* [dks] (continued)

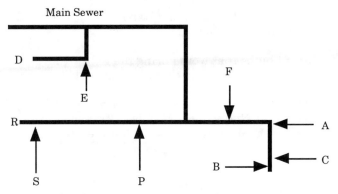

Main Sewer

Figure 4. Sewer map for soap manufacturing facility.

Table 2
**Waste Composition of Soap Manufacturing Facility
Waste Streams**

Waste Source	Sampling Location	COD (mg/L)	BOD (mg/L)	TSS (mg/L)	ABS (mg/L)	Flow (gpm)
Liquid soap	D	1,100	565	195	28	300
Toilet articles	E	2,680	1,540	810	69	50
Soap production	R	29	16	39	2	30
ABS production	S	1,440	380	309	600	110
Powerhouse	P	66	10	50	0	550
Condenser	C	59	21	24	0	1,100
Spent caustic	B	30,000	10,000	563	5	2
Tank bottoms	A	120,000	150,000	426	20	1.5
Fly ash	F	-	-	6,750	-	10
Main sewer		420	260	120	37	2,150

18. *Pollution Prevention Problem 18.* (source reduction, electrical consumption). [jr]. Reducing or eliminating the source of pollution is the most effective method of P^2 available to the engineer or scientist.

A chemistry lab occupies a total of 3,360 ft^2 and is lighted by 36 four-bulb fluorescent fixtures. Each bulb is rated at 45 watts. On the average, the lights are on 8 hours per day, 250 days per year. A published report from the local utility indicates that the local coal-fired electric power plant emits 0.0175 lb of sulfur dioxide, 0.00824 lb of nitrogen oxides, and 2.25 lb of carbon dioxide per kilowatt-hour generated.

It has been suggested that, in the interest of both energy conservation and pollution source reduction, a lower-wattage bulb be used to keep the watts per square foot to a maximum of 1.50, a value that is considered more than adequate for good laboratory lighting.

18. *Pollution Prevention Problem 18.* [jr] (continued)

    a.   Calculate the chemistry lab's average annual lighting energy usage in kW-h/yr.

    b.   Determine the average wattage of the replacement bulbs that would bring the lighting level to 1.50 W/ft^2, assuming that all of the old bulbs are replaced.

    c.   On an annual basis, how many pounds of SO_2, CO_2 and NO_x would NOT be introduced to the atmosphere as a result of the bulb replacement?

III. SOLUTIONS

1. *Pollution Prevention Solution 1.* [sxl]. When shopping, you can do the following things to use fewer resources, produce less waste, and "save" the environment:

- Shop for products that have as little packaging as possible.
- Buy products that use recycled materials in their manufacturing or packaging.
- In any store, do not use a bag unless needed. Have different size cloth bags that can be used and reused for grocery shopping.
- Avoid driving a car unless needed. Car-pool whenever possible.
- When going shopping, plan ahead and make a list of items/materials to be purchased. Reduce the frequency of shopping.
- Do not buy paper towels and napkins; invest in cloth napkins and wiping cloths.

2. *Pollution Prevention Solution 2.* [car]. The possible benefits that should be included in the presentation to Company XYZ management are:

- Reduced operating costs.
- Improved worker safety.
- Reduced compliance costs.
- Increased productivity.
- Improved product quality.
- Increased environmental protection.
- Reduction of future liability costs.
- Continuous improvement, as part of a Total Quality Management (TQM) program.
- Improved public image and relations.

3. *Pollution Prevention Solution 3.* [sn]. The general steps involved in conducting a waste minimization assessment are outlined by the U.S. EPA as follows:

a. Planning and organization
- Get management commitment
- Set overall assessment program goals
- Organize assessment program task force
b. Assessment phase
- Collect process and facility data
- Prioritize and select assessment targets
- Select people for assessment teams

3. *Pollution Prevention Solution 3.* [sn] (continued)

- Generate options
- Screen and select options for further study
  c. Feasibility analysis phase
    - Technical evaluation
    - Economic evaluation
    - Select options for implementation
  d. Implementation
    - Justify projects and obtain funding
    - Equipment installation
    - Implementation (procedure)
    - Evaluation of performance

4. *Pollution Prevention Solution 4.* [jr]. This is an open-ended problem. The following lists are not intended to be comprehensive, but are intended to reflect the type of answers that are reasonable for this question.

   a. Advantages of on-site recovery
   - less waste leaves the facility
   - lower cost of reclaimed solvent (possibly)
   - lower liability for the waste (since it stays at the facility)
   - better control of the quality of the reclaimed solvent
   - less paperwork

   b. Disadvantages of on-site recovery
   - higher capital investment (for recycling equipment)
   - operating costs associated with the recycling
   - higher liability associated with recycling, e.g., accidents, fires, etc.
   - costs for training operators

   c. Advantages of off-site recovery
   - better economics for the recycle processing (since several wastes from different sources are handled at the off-site facility)
   - more outlets for the reclaimed solvents (since other companies are involved)
   - better technical support

   d. Disadvantages of off-site recovery
   - more waste leaves the facility
   - little control of the fate of the waste
   - higher liability associated with the fate of the waste

5. *Pollution Prevention Solution 5.* [sxl]. Suppose that there were 100 lb of discards in the landfill before the paper waste reduction campaign began, and the paper waste was 41 lb of this original total.

Because of the campaign, the paper waste is reduced to 30.75 lb (25% decrease). The new weight percent averages are:

Total weight of discards = 30.75 + 18 + 8 + 9 + 9 + 8 + 7
= 89.75 lb

paper waste = 30.75/89.75 = 0.343 = 34.3 wt%
yard waste = 18/89.75 = 0.201 = 20.1 wt%
glass = 8/89.75 = 0.089 = 8.9 wt%
metals = 9/89.75 = 0.100 = 10.0 wt%
plastic = 9/89.75 = 0.100 = 10.0 wt%
food waste = 8/89.75 = 0.089 = 8.9 wt%
other = 7/89.75 = 0.078 = 7.8 wt%

6. *Pollution Prevention Solution 6.* [sn]. The answer can be tabulated as follows:

Table 3
Plastic Container Symbol Guide with Examples of Container Types

Symbol	Type of Plastic	Examples of container
1-PET	Polyethylene Terephthalate	Beverage bottles, frozen food boil-in-bag pouches, microwave food trays
2-HDPE	High Density Polyethylene	Milk jugs, trash bags, detergent bottles, bleach bottles, aspirin bottles
3-V	Vinyl	Cooking oil bottles, meat packaging
4-LDPE	Low Density Polyethylene	Grocery store produce bags, bread bags, food wraps, squeeze bottles
5-PP	Polypropylene	Yogurt containers, shampoo bottles, straws, syrup bottles, margarine tubs
6-PS	Polystyrene (Styrofoam)	Hot beverage cups, fast food clam-shell containers, egg cartons, meat trays

7. *Pollution Prevention Solution 7.* [wma]. An important, but perhaps not obvious environmental aspect of this product is that juice pack technology has deployed environmentally beneficial technology through such features as:

- Replacement of relatively heavy and "odd-shaped" juice bottles with light-weight, easily stacked containers, resulting in much more efficient transportation and reductions in associated environmental impacts of transport

7. *Pollution Prevention Solution 7.* [wma] (continued)

- Eliminating the need for refrigeration of juices during transport and storage, thereby saving energy, the need for coolants, etc.

Factors associated with the life-cycle analysis of the orange juice include:

- Where and how were the oranges grown? Did the orange grower employ sustainable agricultural practices, e.g., natural pesticides, appropriate fertilizers, etc.?
- Were oranges transported and processed into juice using environmentally responsible methods?
- Were process residues, e.g., pulp, seeds, skins, etc., beneficially used?

Factors associated with the life-cycle analysis of the packaging include:

- Cardboard/Paper:
  - Was recycled paper or virgin pulp from trees used to make the container? Were trees from an "old growth forest" or tree farm used in the production process?
  - Did the paper mill operate in an environmentally responsible manner?
  - What materials were used to manufacture the inks used to print the package? Was ink produced and was printing performed with environmental responsibility?
- Aluminum foil:
  - Was aluminum produced from recycled aluminum, or was "virgin" aluminum used and produced via bauxite ore, alumina, smelting, etc.? (If the source was not recycled aluminum, the environmental impacts of aluminum production must be identified.)
- Plastic:
  - What type of plastic was used, and where did it originate? (To be comprehensive in this analysis, one should identify the environmental impacts of plastic manufacturing beginning with the extraction of crude oil.)

(NOTE: Readers should examine a juice pack to identify all raw materials used in its manufacture.)

8. *Pollution Prevention Solution 8.* [dj]. First, generate an equation for C_1, the total cost as a function of the number of drinks in paper cups:

$$C_1 = \$1.15 \, (N) \tag{1}$$

where N is the number of drinks purchased.

Second, generate an equation for C_2, the total cost as a function of the number of drinks purchased in the plastic cup:

$$C_2 = \$3.18 + \$0.94 \, (N - 1) \tag{2}$$

where N - 1 is used because the first drink in the plastic cup is included in the $3.18 purchase price.

At the break-even point, $C_1 = C_2$, so we obtain:

$$\$1.15 \, (N) = \$3.18 + \$0.94 \, (N - 1) \tag{3}$$

Equation 3 becomes:

$$\$1.15 \, (N) - \$0.94 \, (N) = \$3.18 - \$0.94 \tag{4}$$

Solving both sides of Equation 4 yields:

$$\$0.21 \, (N) = \$2.24 \tag{5}$$

Solving for N yields:

$$N = \$2.24/\$0.21 = 10.67 \tag{6}$$

Since a fraction of a drink cannot be purchased, the break-even point is rounded up to the nearest integer value, which is 11.

The break-even point then is 11 drinks. At this point the initial investment in the reusable plastic cup has been recovered and any additional uses of the cup will make money compared to the use of disposable waxed paper cups.

9. *Pollution Prevention Solution 9.* [dj]. The total cost to the break-even point is:

$$C_2 = \$3.18 + \$0.94 \, (N)$$

where N = 11. Substituting for N yields:

$$C_2 = \$3.18 + \$0.94 \, (11) = \$13.52$$

10. *Pollution Prevention Solution 10.* [dj]. To answer this question, C_1 and C_2 are calculated for N=1,000 uses. The difference,

10. *Pollution Prevention Solution 10.* [dj] (continued)

$C_1 - C_2$, is the amount of money saved:

$$C_1 = \$1.15 \ (1,000) = \$1,150.00$$

$$C_2 = \$3.18 + \$0.94 \ (999) = \$942.24$$

$$C_1 - C_2 = \$1,150.00 - \$942.24 = \$207.76$$

A total of $207.76 was saved over the cost of drinks in disposable paper cups by investing in the reusable plastic cup and buying refills.

11. *Pollution Prevention Solution 11.* [sxl]. To answer this question, the total mass of paper cups thrown away, M_1, and the total crushed volume of paper cups thrown away, V_1, are calculated using equations similar to those shown in Solution 10.

For mass:

$$M_1 = 10 \ \text{g/cup} \ (1,000 \ \text{cups}) = 10,000 \ \text{g}$$

Converting to kg:

$$M_1 = 10,000 \ \text{g/}(1,000 \ \text{g/kg}) = 10 \ \text{kg}$$

For volume:

$$V_1 = 50 \ \text{cm}^3/ \ \text{cup} \ (1,000 \ \text{cups}) = 50,000 \ \text{cm}^3$$

Converting to m³:

$$V_1 = 50,000 \ \text{cm}^3/(1,000,000 \ \text{cm}^3/\text{m}^3) = 0.050 \ \text{m}^3$$

12. *Pollution Prevention Solution 12.* [dj]. To solve this problem, an energy cost for the paper cups, E_1, is first calculated using the equation:

$$E_1 = 400,000 \ \text{joule/cup} \ (N) \tag{1}$$

where N is the number of cups. The break-even value, N = 11, calculated in Solution 8 is used for the calculations in this problem. Substituting into Equation 1 yields:

$$E_1 = 400,000 \ \text{joule/cup} \ (11 \ \text{cups}) = 4,400,000 \ \text{joules} \tag{2}$$

To calculate the quantity of oil used, E_1 is divided by the energy content in a gallon of oil as follows:

12. *Pollution Prevention Solution 12*. [dj] (continued)

Oil volume = (E_1)/(Energy per gallon of oil)

Oil volume = (4,400,000 joules)/(142,000,000 joules/gallon)
= 0.031 gallons.

This quantity is:

(0.031 gallons) (128 fluid ounces/gallon)
= 4.0 fluid ounces = 117 mL

about the size of a small glass of orange juice.

13. *Pollution Prevention 13*. [dj]. To solve this problem, the energy cost for the fabrication and use of the plastic cup is calculated using the equation:

$$E_2 = 800,000 \text{ joules} + 394,000 \text{ joule/cup (N)} \tag{1}$$

where N, and not N - 1, is used because the plastic cup must be washed every time it is used. Substituting the break-even value, N = 11, that was calculated in Solution 8 into Equation 1 yields:

$$E_2 = 800,000 \text{ joules} + 394,000 \text{ joule/cup (11 cups)}$$
$$= 5,134,000 \text{ joules} \tag{2}$$

Since E_2 exceeds E_1 = 4,400,000 joules (from Solution 12), then, at the cost break-even point, more energy was used to make and wash the plastic cup than was used to make and throw away 11 waxed paper cups.

14. *Pollution Prevention Solution 14*. [dj]. To calculate the energy break-even point, methods similar to those used in Solution 8 are utilized here. Equations are written for E_1 and E_2, they are set equal to each other, and they are solved for N. From Solution 12:

$$E_1 = 400,000 \text{ joule/cup (N)} \tag{1}$$

From Solution 13:

$$E_2 = 800,000 \text{ joules} + 394,000 \text{ joule/cup (N)} \tag{2}$$

Equating the right-hand sides of Equations 1 and 2 yields:

$$400,000 \text{ joule/cup (N)} = 800,000 \text{ joules} + 394,000 \text{ joule/cup (N)} \tag{3}$$

14. *Pollution Prevention Solution 14.* [dj] (continued)

Solving Equation 3 yields:

$$6,000 \text{ joule/cup (N)} = 800,000 \text{ joules}$$

$$N = (800,000 \text{ joules})/(6,000 \text{ joule/cup}) = 133.33 \text{ cups}$$

Since, again, fractional drinks cannot be purchased the energy break-even point is rounded up to the nearest integer value, N = 134. At values of N greater than 134, buying drinks for the reusable plastic cup will use less total energy than the amount used to buy drinks in throw-away paper cups.

15. *Pollution Prevention Solution 15.* [dj]. To solve this problem, the energy savings over the life of one plastic cup must first be calculated. For N = 1000, the energy requirements of the wax paper and reusable cups, E_1 and E_2, respectively are:

$$E_1 = 400,000 \text{ joule/cup } (1,000) = 400,000,000 \text{ joules}$$

$$E_2 = 800,000 \text{ joules} + 394,000 \text{ joule/cup } (1,000) = 394,800,000 \text{ joules}$$

Taking the difference yields the energy saved over the life of the reusable plastic cup, or:

$$E_1 - E_2 = 400,000,000 - 394,800,000 = 5,200,000 \text{ joules}$$

If oil is used at 40% efficiency for the energy needed to make and wash the plastic cup, then the energy yield obtained from a gallon of oil is:

$$\text{Yield} = (40/100) \ (142,000,000 \text{ joules/gallon})$$
$$= 56,800,000 \text{ joules/gallon}$$

Therefore, the oil saved over the life of one cup is the energy savings divided by the energy yield, or:

$$\text{Oil saved per cup} = (5,200,000 \text{ joules})/(56,800,000 \text{ joules/gallon})$$
$$= 0.092 \text{ gallons}$$

To calculate the oil saved for 1,000,000 cups, the oil saved per cup is multiplied by 1,000,000, yielding a total oil volume saved of:

$$\text{Total oil saved} = (\text{oil saved per cup}) \ (1,000,000 \text{ cups})$$
$$= (0.092 \text{ gallons/cup}) \ (1,000,000 \text{ cups}) = 92,000 \text{ gallons.}$$

This oil savings is sufficient to run approximately 92 cars for a year!

16. *Pollution Prevention Solution 16.* [jr]

   a. The emission factor for the organic solvent-based coating, f_o, is given by:

$$f_o = \frac{x_v \, x_w \, \rho_v}{1 - x_s} = \frac{(1)\,(0.40)\,(7.36)}{(1 - 0.40)} = 4.91 \text{ lb}_{voc}/\text{gal}_{solids}$$

   b. The emission factor for the water solvent-based coating, f_w, is given by:

$$f_w = \frac{x_v \, x_w \, \rho_v}{1 - x_s} = \frac{(0.20)\,(0.65)\,(7.36)}{(1 - 0.65)} = 2.73 \text{ lb}_{voc}/\text{gal}_{solids}$$

   c. The percent reduction in VOC emissions would be:

$$\% \text{ reduction} = \left(\frac{f_o - f_w}{f_o} \right)(100) = \left(\frac{4.91 - 2.73}{4.91} \right)(100) = 44.4\%$$

17. *Pollution Prevention Solution 17.* [dks].

   a. The first step in the solution to this problem is to determine the mass loadings at each location on the sewer map using the table of data given in the problem statement. This is done by multiplying the concentration by the flow rate to yield a mass flow rate, with proper unit conversions. The result is shown in Table 4. Also given in Table 5 are the percentages of the total mass of each contaminant associated with a given sampling location.

   b. A study of the sewer map and the loading data should yield clues about which waste streams would be candidates for segregation. For example, the fly ash line, F, carries 26 wt% of the suspended solids but only 0.5% of the flow. This suggests that segregation of this source would reduce the solids loading significantly at a small cost, since the flow is so small. In fact, it may be possible to eliminate the water from this source directly and treat the fly ash in a dry

17. *Pollution Prevention Solution 17.* [dks] (continued)

Table 4
**Mass Loadings for Soap Manufacturing Facility
Waste Streams**

Waste Source	Sampling Location	COD (lb/d)	BOD (lb/d)	TSS (lb/d)	ABS (lb/d)	Flow (gpm)
Liquid soap	D	3,963	2,036	703	101	300
Toilet articles	E	1,609	925	486	41	50
Soap production	R	10	6	14	1	30
ABS production	S	1,902	502	408	793	110
Powerhouse	P	436	66	330	0	550
Condenser	C	779	277	317	0	1,100
Spent caustic	B	721	240	14	0	2
Tank bottoms	A	2,162	2,702	8	0	1.5
Fly ash	F	0	0	811	0	10
Main sewer		11,619	6,713	3,098	955	2,150

Table 5
**Percent Mass Loadings for Each Waste Stream
in the Soap Manufacturing Facility**

Waste Source	Sampling Location	COD (wt%)	BOD (wt%))	TSS (wt%)	ABS (wt%)	Flow (gpm)
Liquid soap	D	34.2	30.1	22.7	10.8	13.9
Toilet articles	E	13.9	13.7	15.7	4.4	2.3
Soap production	R	0.1	0.1	0.5	0.1	1.4
ABS production	S	16.4	7.4	13.2	84.7	5.1
Powerhouse	P	3.8	1.0	10.7	0	25.5
Condenser	C	6.7	4.1	10.3	0	51.1
Spent caustic	B	6.2	3.6	0.4	0	0.1
Tank bottoms	A	18.7	40.0	0.2	0	0.1
Fly ash	F	0	0	26.2	0	0.5
Main sewer		100	100	100	100	100

state. Similarly the tank bottoms contain a significant amount of organic matter but very little suspended solids and almost no flow. Segregation here would also be beneficial to eliminate this large source of organic material. In order to help decide which flows could be segregated, two indices can be used.

The first index is calculated based on the product of the mass loading and the square of the concentration for a given source, divided by the product of the mass loading and the square of the concentration in the main sewer. This index emphasizes the higher concentration sources as candidates because generally, higher

17. *Pollution Prevention Solution 17.* [dks] (continued)

concentration wastes are easier to treat. The values of the index for each contaminant from each source is given in Table 6.

Table 6
Concentration Index Based on Main Sewer Composition for Each Waste Stream in the Soap Manufacturing Facility

Waste Source	Sampling Location	COD	BOD	TSS	ABS
Liquid soap	D	2	1	1	0
Toilet articles	E	5	5	7	0
Soap production	R	0	0	0	0
ABS production	S	2	0	1	218
Powerhouse	P	0	0	0	0
Condenser	C	0	0	0	0
Spent caustic	B	276	53	0	0
Tank bottoms	A	13,230	133,970	0	0
Fly ash	F	0	0	828	0

The second index is based not on the concentration in the main sewer, but on typical secondary wastewater treatment standards for each contaminant. The ratio of the actual concentration to the concentration for the secondary standard is used to determine which source has the most impact on the plant related to the plant's ability to meet these standards for pretreatment or direct discharge purposes. This index is shown below in Table 7.

Table 7
Index Based on Secondary Municipal Wastewater Treatment Standards for Each Waste Stream in the Soap Manufacturing Facility

Waste Source	Sampling Location	COD	BOD	TSS	ABS
Liquid soap	D	27	7	30	1
Toilet articles	E	64	24	355	1
Soap production	R	0	0	0	0
ABS production	S	22	1	43	1,902
Powerhouse	P	0	0	1	0
Condenser	C	0	0	0	0
Spent caustic	B	3,603	267	5	0
Tank bottoms	A	172,938	675,540	2	0
Fly ash	F	0	0	41,039	0
Standards		60	30	30	5

17. *Pollution Prevention Solution 17.* [dks] (continued)

In each table, the very large numbers stand out for COD and BOD in the tank bottoms (A) and spent caustic (B), ABS in the ABS production stream (S), and suspended solids in the fly ash (F) line. This shows up consistently for both indices. Additionally, suspended solids in line (E) have a large value in the second index.

These groupings suggest segregation as shown in Figure 5.

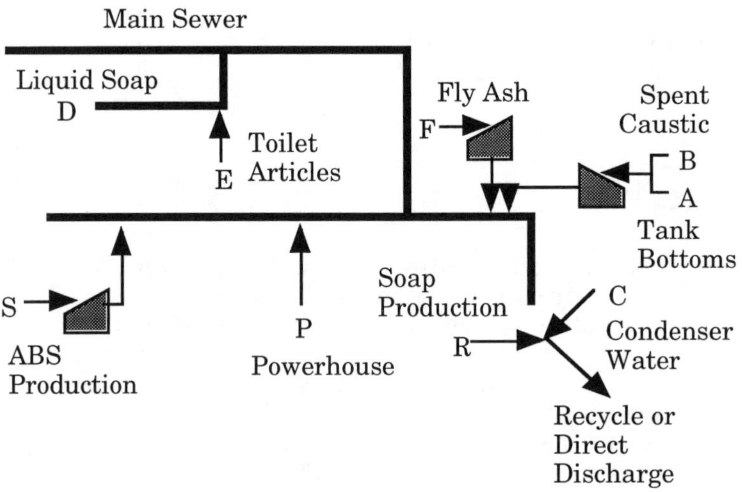

Figure 5. Revised sewer map for soap manufacturing facility showing waste streams identified for consideration for waste segregation and recycling.

The new map takes the lines A,B,F, and S, and provides some pretreatment (say 90%) before discharge to the sewer. Lines R and C are below secondary treatment standards and may be discharged directly to surface water or may be reused in the plant to reduce its overall water demand. The new waste stream composition and mass balance tables reflecting these changes are shown in Tables 8 and 9.

17. *Pollution Prevention Solution 17.* [dks] (continued)

Table 8
Waste Stream Composition for Soap Manufacturing Facility Following
Waste Segregation Analysis as Indicated in Figure 5

Waste Source	Sampling Location	COD (mg/L)	BOD (mg/L)	TSS (mg/L)	ABS (mg/L)	Flow (gpm)
Liquid soap	D	110	56	19	28	300
Toilet articles	E	268	154	81	69	50
Soap production	R	29	16	39	2	30
ABS production	S	720	190	155	60	110
Powerhouse	P	7	1	5	0	550
Condenser	C	59	21	24	0	1,100
Spent caustic	B	3,000	1,000	56	5	2
Tank bottoms	A	12,000	15,000	43	20	1.5
Fly ash	F	0	0	675	0	10
Main sewer		450	260	120	37	2,150

Table 9
Mass Loadings for Soap Manufacturing Facility Waste Streams Following
Waste Segregation Analysis as Indicated in Figure 5

Waste Source	Sampling Location	COD (lb/d)	BOD (lb/d)	TSS (lb/d)	ABS (lb/d)	Flow (gpm)
Liquid soap	D	396	204	70	10	300
Toilet articles	E	161	92	49	41	50
Soap production	R	10	6	14	1	30
ABS production	S	951	251	204	79	110
Powerhouse	P	44	7	33	0	550
Condenser	C	779	277	317	0	1,100
Spent caustic	B	72	24	1	0	2
Tank bottoms	A	216	270	1	0	1.5
Fly ash	F	0	0	81	0	10
Main sewer		2,630	1,131	770	223	2,150

A new set of segregation indices were calculated as above and reflect the changes indicated in Figure 5 and Tables 8 and 9. As seen in Tables 10 and 11, further reductions in the COD and BOD in Line A may be necessary, or it may be appropriate to isolate Line A for separate treatment by virtue of its low flow rate and very high BOD and COD concentrations, even after 90% reduction following treatment.

17. *Pollution Prevention Solution 17.* [dks] (continued)

Table 10
Concentration Index Based on Main Sewer Composition for Each Waste
Stream in the Soap Manufacturing Facility Following Waste Segregation
Indicated in Figure 5

Waste Source	Sampling Location	COD	BOD	TSS	ABS
Liquid soap	D	0	0	0	1
Toilet articles	E	0	1	1	3
Soap production	R	0	0	0	0
ABS production	S	12	2	9	4
Powerhouse	P	0	0	0	0
Condenser	C	0	0	0	0
Spent caustic	B	16	6	0	0
Tank bottoms	A	755	15,070	0	0
Fly ash	F	0	0	66	0

Table 11
Index Based on Secondary Municipal Wastewater Treatment Standards for
Each Waste Stream in the Soap Manufacturing Facility Following Waste
Segregation Indicated in Figure 5

Waste Source	Sampling Location	COD	BOD	TSS	ABS
Liquid soap	D	0	0	0	53
Toilet articles	E	1	0	4	132
Soap production	R	0	0	0	0
ABS production	S	27	1	54	190
Powerhouse	P	0	0	0	0
Condenser	C	0	0	2	0
Spent caustic	B	36	3	0	0
Tank bottoms	A	1,729	6,755	0	0
Fly ash	F	0	0	410	0
Standards		60	30	30	5

18. *Pollution Prevention Solution 18.* [jr].

a. The total electrical (lighting) power, P, for the
lab is

$$P = (36 \text{ fixtures}) (4 \text{ bulbs/fixture}) (45 \text{ W/bulb}) = 6{,}480 \text{ W}$$

The number of hours per year, t, that the lights are
on is:

$$t = (250 \text{ d/yr}) (8 \text{ hr/d}) = 2{,}000 \text{ hr/yr}$$

The annual electrical energy usage for lighting, E,
is:

18. *Pollution Prevention Solution 18.* [jr] (continued)

$E = (6{,}480 \text{ W}) (0.001 \text{ kW/W}) (2{,}000 \text{ hr/yr}) = 12{,}960 \text{ kW-hr/yr}$

b.   The present lighting level in the lab is:

$(6{,}480 \text{ W})/(3{,}360 \text{ ft}^2) = 1.929 \text{ W/ft}^2$

To bring the level down to 1.5 W/ft², the wattage should be multiplied by the ratio, 1.5/1.929. Therefore:

recommended bulb wattage $= (1.5/1.929) (45 \text{ W/bulb})$
$= 35 \text{ W/bulb}$

c.   Using the replacement bulbs, the savings in annual energy usage would be

$(12{,}960 \text{ kW-hr/yr}) (1 - 1.5/1.929) = 2{,}882 \text{ kW-hr/yr}$

The reduction of pollutants from the utility would therefore be:

SO_2: $(0.0175 \text{ lb/kW-hr}) (2{,}882 \text{ kW-hr/yr}) = 50.4 \text{ lb/yr}$

CO_2: $(2.25 \text{ lb/kW-hr}) (2{,}882 \text{ kW-hr/yr}) = 6{,}485 \text{ lb/yr}$

NO_x: $(0.00824 \text{ lb/kW-hr}) (2{,}882 \text{ kW-hr/yr}) = 23.8 \text{ lb/yr}$

Chapter 5

ENERGY CONSERVATION

Gary Hickernell

I. INTRODUCTION

Many of us still remember the fuel shortages of the 1970s with long lines at filling stations, and buildings under-heated or closed in the winter. Others remember, more nostalgically, gasoline prices of 32 cents per gallon and spending the evening driving around on a dollar's worth of gas. Gasoline at that price is long gone, probably never to return, and a return to the days of fuel shortages may be lurking just around the corner. In our highly industrialized society, Americans are by far the largest energy users in the world and must seek ways to conserve energy. Small percentage improvements in the efficiency of energy use translate into large amounts of coal, fuel oil, and gasoline made available for other uses or that may be banked for future generations. Such increased efficiency also reduces the production of air pollutants such as the greenhouse gas, carbon dioxide, that contributes to global warming. Emissions of sulfur and nitrogen oxides which are involved in the production of acid rain are also reduced when energy is conserved.

Whatever our reasons, energy conservation should be an important part of our pollution control strategies. There are many things, both small and inexpensive and large and costly, that can be done to improve the manner in which we use this valuable resource. The May 1997 issue of the magazine *Popular Science* (Stover, 1997) reported on efforts to design housing that reduces energy use by 50% without increasing the per square foot cost. This meets the overall goal of matching a practical solution with short-term improvements in energy efficiency. Approaches to energy conservation in industrial and institutional settings need also to have this same goal.

Transportation is by far the largest single energy-consuming sector of the economy. The U.S. has made some strides in improving the fuel efficiency of automobiles, buses, trucks and airplanes, although recently there has been a trend away from fuel efficient cars. This is a potential problem, since due to the volume of fuel consumed, small changes in fuel efficiency produce very large changes in national fuel consumption. Fundamental changes in the way we travel, such as increased use of mass transit or extensive carpooling, could contribute significantly to energy conservation. However, such solutions are beyond the realm of engineering and science, and must be embraced by our society as a whole.

More immediate to our fuel supply future are the engineering changes that can be made, particularly in energy use by municipalities, institutions and industry. In these contained environments, societal issues can be kept at bay and energy conservation can be seen in the context of return on investment and environmental compliance, both very real and concrete issues.

Environmental management is, to a large degree, focused on the extraction, production, use and management of energy resources. It is worthwhile considering all choices, the context in which the choices are made, and their potential influence on the bigger picture. Knee-jerk responses to energy management problems often result in poor choices and often are not sustainable over the long run.

This chapter is designed to further the understanding of energy conservation conceptually and to quantify the degree to which energy conservation can be achieved in some specific cases. Although the exercises provided in this chapter are not difficult from a technical perspective, it is hoped that they will provide the student with the opportunity to consider choices in energy and to assess the effect of those choices on resource utilization and the environment.

Reference: Stover, D., Reengineering the American home, *Popular Science*, 125(5), 62-65, 1997.

II. PROBLEMS

1. *Energy Conservation Problem 1.* (conservation, cost savings, energy, illumination, payback period). [dj]. Ivory Tower University runs its own coal-fired power plant, consuming Utah bituminous coal with an energy content (in the combustion literature, energy content is defined as the lower heating value, LHV) of 25,000 kilojoules/kg. The coal contains, on average, 1.0 wt% sulfur and 1.2 wt% ash (based on the total mass of the coal). The power plant is 35% efficient (meaning that 35% of the energy in the coal is actually converted to electrical energy), and is operated at a 2.0 megawatt average daily electrical load (ADL). Assume that the coal is completely burned during combustion, and also that the power plant captures 99% of the ash and 70% of the sulfur dioxide produced during combustion. The current coal price is $120/ton delivered to the university. Ash hauling charges to the regional landfill are $40/metric tonne (1 metric tonne = 1,000 kg = 2,205 pounds = 1.1025 U.S. tons).

After a U.S. EPA Green Lights energy audit, Ivory Tower finds that it can install energy-efficient lighting and reduce its average daily electrical generating needs by 25%. The materials and labor costs for the energy-efficient lighting upgrades are $350,000, which Ivory Tower will pay from cash on hand.

Using the information given above, calculate the average reduction in electrical load, and the new average daily load for the power plant.

2. *Energy Conservation Problem 2.* (conservation, cost savings, energy, illumination, payback period). [dj]. Refer to Problem 1. Using the efficiency of the power plant, the heating value of the coal, and the results from Problem 1, calculate the daily <u>reduction</u> in the quantity of coal (kg/d) consumed by the university's power plant.

3. *Energy Conservation Problem 3.* (conservation, cost savings, energy, illumination, payback period). [dj]. Using the results from Problem 2, the ash content of the coal and the ash removal efficiency from Problem Statement 1, calculate the daily <u>reduction</u> in the quantity of ash (kg/d) produced when the university implements this energy-saving lighting program.

4. *Energy Conservation Problem 4.* (conservation, cost savings, energy, illumination, payback period). [dj]. Using the results from Problem 2, and the quoted price of the coal in Problem Statement 1, calculate the annual reduction in cost of coal supplied to the power plant when the university implements this energy-saving lighting program.

5. *Energy Conservation Problem 5.* (conservation, cost savings, energy, illumination, payback period). [dj]. Using the results from Problem 3 and the quoted price for disposal of the ash in Problem Statement 1, calculate the annual reduction in cost of ash hauled to the landfill when the university implements this energy-saving lighting program.

6. *Energy Conservation Problem 6.* (conservation, cost savings, energy, illumination, payback period). [dj]. Using the results from Problems 4 and 5 and the quoted cost to install the efficient lighting systems from Problem Statement 1, calculate the time in years required for the cost savings to pay back the cost of installing the efficient lighting at Ivory Tower University.

7. *Energy Conservation Problem 7.* (conservation, cost savings, energy, illumination, payback period). [dj]. Refer to Problem 1. Sulfur dioxide (SO_2) is produced from the coal's sulfur during combustion. The atomic weight of sulfur is 32 kg/kg-mole; the atomic weight of SO_2 is 64 kg/kg-mole. Using this information, the results from Problem 2, the sulfur content of the coal and the power plant removal efficiency from Problem Statement 1, calculate the annual reduction (metric tonnes/year) of SO_2 emissions that will result from the 25% reduction in average daily electrical generating needs.

8. *Energy Conservation Problem 8.* (conservation, energy conversion, efficiency). [dks]. In 1900 it took about 20,000 BTU fuel input to produce 1 kW-h of electricity. Estimate the efficiency of conversion and compare it with a typical value for today's power industry.

9. *Energy Conservation Problem 9.* (conservation, energy conversion, efficiency, thermodynamics). [dks]. A steam turbine in a power plant operates between the temperatures of 300 and 900 K. The efficiency has already been increased to the maximum by mechanical changes, so we are trying to improve the efficiency thermodynamically. We find we can increase the operating temperature of the heat source by 30 K or decrease the temperature of the heat sink (the cool reservoir) by 30 K. Which is most effective? Do you think it would be a significant improvement to use water from the bottom of a lake rather than the top for cooling water?
NOTE: The following sketch provides a definition of efficiency in a hypothetical heat engine, and should be reviewed before completing the problem.

9. *Energy Conservation Problem 9.* [dks] (continued)

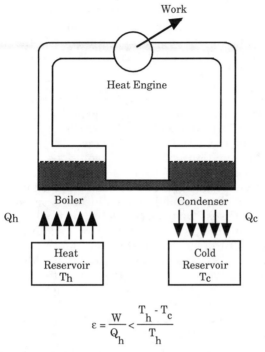

$$\varepsilon = \frac{W}{Q_h} < \frac{T_h - T_c}{T_h}$$

Figure 6. Hypothetical heat engine.

10. *Energy Conservation Problem 10.* (conservation, energy conversion, waste-to-energy, solid waste combustion, thermodynamics, steam production). [dks]. The city of Nashville, TN, experimented with the use of municipal solid waste as a source of energy to heat and air condition downtown buildings. The idea arose when a 1972 law requiring daily earth cover at all landfills was added to the existing problem of having only enough landfill volume for 12 years. The landfill, so habitually used, was no longer a simple and cheap solid waste management solution.

Phase 1 of the waste-to-energy project included a multiple grate incinerator, dry cyclone collector for particulates, and a wet scrubber for acid gas removal. It burned 720 ton/d of municipal solid waste. How many pounds of steam should it have produced daily? (Assume that the municipal solid waste contains 5,000 Btu/lb, and that a total of 1,060 Btu are required to generate 1 lb of steam.)

11. *Energy Conservation Problem 11.* (energy conservation, electricity, coal, nuclear energy, renewable energy). [wma]. Even with an aggressive energy conservation program, the Earth's

11. *Energy Conservation Problem 11.* [wma] (continued)

growing population will continue to demand increasing amounts of electricity. Identify and describe the environmental impacts, both positive and negative, of the following alternative means of power generation: coal-fired steam boilers; nuclear power; photovoltaic solar panels; and hydroelectric dams.

12. *Energy Conservation Problem 12.* (conservation, fuel economy, fuel additives, automobile, transportation). [glh]. As mentioned in the Introduction to Chapter 5, the opportunity for energy conservation in the field of transportation is enormous. Forty-five million gallons of gasoline could be saved for future generations *each day in the U.S. alone* with a small increase in fuel efficiency of only 2 mpg. And, as has been stressed before, this is a "win-win" solution since reduction in pollution is proportional to the reduction in fuel use. One partial solution to the production of combustion pollutants is the addition of gasoline additives to increase the amount of oxygen in the fuel. Methyl-*tert*-butyl ether (MTBE) is currently used as an oxygenating fuel additive, providing more power per gallon as well as more complete gasoline combustion in internal combustion engines. This is achieved at a cost of about 1 cent per gallon of fuel.

Assuming that the average fuel usage rating for automobiles in the United States is 18 mpg and that the increase in efficiency with the MTBE additive is 5%, calculate the savings in gallons of fuel in the United States per year if MTBE were used in every car.

13. *Energy Conservation Problem 13.* (conservation, fuel economy, fuel additives, automobile, transportation). [glh]. An important consideration for the use of the MTBE fuel additive (see Problem 12) is the "break-even" point. This occurs when the cost of the additive equals the savings in gasoline. Given that the use of MTBE costs 1 cent per gallon and the cost of gasoline averages at $1.20 per gallon nationwide, calculate the "break-even" point for its use. Use any relevant data from Problem 12 in the solution to this problem.

14. *Energy Conservation Problem 14.* (conservation, fuel economy, fuel additives, automobile, transportation). [glh]. Discuss the use of the MTBE fuel additive in light of the results of Problem 11. Two additional factors have been recently added to the arguments against MTBE addition. First, MTBE is thought to be a possible health threat to consumers as they pump their own gas, and secondly, MTBE is highly mobile in groundwater, has generated large groundwater plumes below a number of leaking underground storage tank sites, and poses a potential health threat due to contamination of drinking water supplies.

15. *Energy Conservation Problem 15*. (conservation, recycling, life cycle analysis, LCA). [glh]. A good plan for energy conservation will include a variety of strategies. One of these should be recycling of materials. Most recycling efforts have multiple positive benefits, including reducing energy demands and landfill space, as well as reducing the pollution of land, air and water.

Our society continues in its transition from a "throw-away" society to one which is giving more thought to each consumer product and its containers. This change has been brought about by a variety of pressures, including dwindling landfill space and the increased cost of solid waste disposal. Incentives in the form of bottle deposits and recycling efforts of communities have helped in this transition.

The engineers at Quality Container Corporation have designed three new containers and are making a presentation to the Vice President of Marketing. She will make the choice of which container to use to package a new product. Differences in appearance of the three options are negligible and none has an effect on the product in any way.

Using the data below, help the Vice President make the energy-conscious decision with regard to this container.

	Container A	Container B	Container C
E_1	200 J/unit	450 J/unit	650 J/unit
E_2	0 J/unit	250 J/unit	150 J/unit

where E_1 = energy to produce the container, and E_2 = energy to manufacture the container from recycled materials.

Assume that the raw materials cost to produce the item are the same for each container. The only difference is in the amount of energy needed to improve the structure of the container to make it more durable. Container A cannot be recycled while containers B and C can be recycled. However, Container B can only be recycled five times, on average, while Container C can be recycled up to 20 times without deterioration of performance.

a. Make a table for each container showing the number of units required for each recycle and the associated costs, E_1, E_2 and Total E. Calculate the energy costs for one million unit-uses at a recycle rate of 25%.

b. Which container would you advise the Vice President to select, based only upon these energy considerations?

16. *Energy Conservation Problem 16.* (conservation, recycling, life cycle analysis, LCA). [glh]. Using the data from Problem 15, calculate the number of times Containers B and C would have to be recycled to make them competitive with Container A, assuming a recycle rate (RR) of 90%.

III. SOLUTIONS

1. *Energy Conservation Solution 1.* [dj]. With a 25% reduction in electrical load resulting from the implementation of energy conservation measures, the new electrical load will be 100% - 25% = 75% of the old electrical load. For a 2.0 mW power plant, the new average daily load, ADL will be:

New ADL = (Old load) (100% - % savings)/100%
= (2.0 mW) (100% - 25%)/100% = (2.0 mW) (0.75) = 1.5 mW

The average reduction in electrical load will be:

average reduction = (new ADL) - (old ADL)
= (2.0 mW) - (1.5 mW) = 0.5 mW

2. *Energy Conservation Solution 2.* [dj]. First, use the available data to calculate the thermal energy input to the plant. Recall that 1 watt = 1 joule/s, 1 kilojoule = 1.0×10^3 joules, and 1 mW = 1.0×10^6 watts. The energy contained in one joule can raise the temperature of 1 gram of water by 0.25 °C.

Before energy conservation:

Thermal energy input = (Electrical output)/(fractional thermal efficiency)

where fractional thermal efficiency is expressed as a decimal.

For a 35% efficient power plant, the fractional thermal efficiency is:

35%/100% = 0.35

Therefore,

Thermal energy input = $(2.0 \times 10^6$ joule/s)/(0.35) = 5.7×10^6 joule/s

After energy conservation, the same equation is used to calculate thermal energy input, but the new, lower electrical energy requirement is used:

Thermal energy input = $(1.5 \times 10^3$ joule/s)/(0.35) = 4.3×10^3 joule/s

Next, use the energy content of the coal (called the lower heating value, LHV) to calculate the mass of coal required for the thermal energy input.

Before energy conservation:

2. *Energy Conservation Solution 2.* [dj] (continued)

> Mass rate of coal = (Thermal energy input)/(LHV of coal)
> Mass rate of coal = $(5.7 \times 10^6$ joule/s)/$(25.0 \times 10^6$ joule/kg)
> = 0.23 kg/s

After energy conservation:

> Mass rate of coal = $(4.3 \times 10^6$ joule/s)/$(25.0 \times 10^6$ joule/kg)
> = 0.17 kg/s

The reduction in mass rate of coal used is:

> Before mass rate - After mass rate = 0.23 kg/s - 0.17 kg/s
> = 0.06 kg/s

Converting to a daily reduction:

> Mass coal/ d = (mass per second) (number of seconds per day)
> Mass coal/d = (0.06 kg/s) (86,400 s/d) = 5,200 kg/d

3. *Energy Conservation Solution 3.* [dj]. To get the fractional capture efficiency of the power plant, divide the percentage captured by 100:

> 99/100 = 0.99 kg ash captured/1.00 kg ash in coal

To get the mass fraction of ash in the coal, divide the percent ash by 100:

> 1.2/100 = 0.012 kg ash/kg coal

To calculate the daily reduction in captured ash produced by the power plant, multiply the mass of coal per day by the mass fraction of ash in the coal, and multiply the result by the fractional capture efficiency of the power plant:

> (Mass coal/d) (mass ash/mass coal) (capture efficiency)
> = (5,200 kg coal/d) (0.012 kg ash/kg coal) (0.99 kg captured/kg ash)
> = 62 kg ash captured/d

Note that since 5,200 kg coal is the reduction in the amount of coal needed per day, then the 62 kg/d result is the reduction in the amount of ash that is caught per day.

4. *Energy Conservation Solution 4.* [dj]. To calculate the annual reduction in the cost of coal supplied to the power plant, multiply the daily savings in coal by the number of days in a year, then multiply the result by the price of coal in dollars per kg:

4. *Energy Conservation Solution 4.* [dj] (continued)

Annual cost in $/yr = (mass coal in kg/yr) (price of coal in $/kg)

mass of coal in kg/year = (5,200 kg/d) (365 d/yr)
= 1,898,000 kg/yr

price of coal in $/kg = ($120/ton)/(1,000 kg/1.1025 ton)
= $0.13/kg

Substituting these results yields:

Annual cost of coal = (1,898,000 kg/year) ($0.13/kg)
= $248,000/yr

Since the $248,000/yr figure is based on a savings of 5,200 kg coal/d, then the $248,000 figure represents the annual cost savings for coal consumed by the power plant.

5. *Energy Conservation Solution 5.* [dj]. To calculate the annual reduction in ash hauling costs, multiply the daily reduction in ash by the number of days per year, then multiply the result by the cost of hauling ash to the landfill.

Annual ash savings in $/yr
= (mass ash in kg/d) (365 d/yr) (ash haul price in $/kg)
= (62 kg/d) (365 d/yr) ($40/tonne) (1 tonne/1,000 kg)
= $905/yr

6. *Energy Conservation Solution 6.* [dj]. To calculate the time to pay back the cash investment to upgrade the lighting system, first calculate the total annual savings by adding up the savings for the cost of the coal and the cost of hauling the ash:

The total annual savings are = $248,000/yr + $905/yr
= $248,905/yr

Next, divide the quoted upgrade cost by the annual savings:

Time to recover cost = (Upgrade cost in $)/(Annual savings in $/yr)
= ($350,000)/($248,905/yr) = 1.4 yr

This represents a very rapid pay-back period in addition to producing lower levels of CO_2, ash, and SO_2 pollution from the reduction in energy use, and the installation of this more efficient lighting system should proceed immediately.

7. *Energy Conservation Solution 7.* [dj]. The reduction in SO_2 emissions from the power plant is proportional to the reduction in

7. *Energy Conservation Solution 7.* [dj] (continued)

mass of coal used. First calculate the reduction in mass of sulfur produced per day:

Mass S in kg/d
= (mass reduction coal, kg/d) (fraction sulfur in coal, kg S/kg coal)
= (5,200 kg/d) (0.01 kg S/kg coal) = 52 kg S/d

Next, convert the reduction in sulfur mass to reduction in sulfur dioxide mass using the ratio of the atomic weights of sulfur dioxide to sulfur:

Mass SO_2 in kg/d = (mass S in kg/d) (MW SO_2)/(AW S)
= (52 kg S/d) (64 kg SO_2/kg-mole)/(32 kg S/kg-mole)
= (52 kg S/d) (2 kg SO_2/ kg S) = 104 kg SO_2/d

This is actually the daily mass of SO_2 that is not generated when we saved 5,200 kg coal/d.

Next, calculate the mass of SO_2 that would be emitted from the power plant's stack by multiplying the mass of SO_2 not generated by (1 - percent collection efficiency/100). Remember that the power plant has scrubbing equipment that removes 70% of the SO_2 from its exhaust stream:

Mass SO_2 saved from release in kg/d
= (1 - 70/100) (mass SO_2 not generated in kg/d)
= (0.30) (mass SO_2 not generated in kg/d)
= (0.30) (104 kg SO_2/d) = 31.2 kg SO2/d

The value of 31.2 kg SO_2/d is the daily mass that is not released to the atmosphere.

Finally, calculate the annual release reduction in tonnes/year by multiplying the daily release reduction by the number of days in a year and dividing by the number of kg in a metric tonne:

Mass SO_2 saved in tonne/yr
= (31.2 kg SO_2/d) (365 d/yr) (1 tonne/1,000 kg)
= 11.4 tonne/yr

8. *Energy Conservation Solution 8.* [dks]. The solution to this problem is based on unit conversions. From standard conversion tables, 1 kW-h is equivalent to 3,412 Btu. Since only 1 kW-h was being produced, the efficiency of energy conversion is:

(20,000 Btu)/(3,412 Btu/kW-h) = 5.86 kW-h

Since only 1 kW-h was being produced, the efficiency of energy conversion is:

8. *Energy Conservation Solution 8.* [dks] (continued)

Efficiency = (actual energy produced)/(energy production potential)
        = (1 kW-h)/(5.85 kW-h) = 0.171 = 17.1%

Today's energy conversion efficiency has improved over this value from 96 years ago (thankfully!) but not as much as we would like. Typical values range from 30 to 35% efficiency, or 100% better than before. However, as you can see, there is much room for improvement!

9. *Energy Conservation Solution 9.* [dks]. Using the definition of efficiency as indicated in the figure in the problem statement, the impact on power generation efficiency of changing the system operating temperatures can be evaluated. If the temperature of the heat source, T_h, is changed, the numerator of the efficiency equation increases but the denominator increases as well, so the increase in energy generation efficiency is small. Using a numerical example:

Efficiency = $(T_h - T_c)/T_h$ = (930 - 300)/(930) = 0.677

versus the initial value of:

Efficiency = (900 - 300)/(900) = 0.667

or a 1.5% increase. Alternatively, by cooling the cooling water, the only impact is in the numerator while the denominator stays the same. Numerically:

Efficiency = (900 - 270)/(900) = 0.70

for a 5% increase in efficiency. So, cooling the heat sink does more to improve efficiency than raising the temperature of the hot side of the heat engine.

The water at the bottom of the lake would, generally, be cooler than at the top. If the cost of obtaining water from the bottom were the same as at the top, then the improvement in turbine efficiency may be significant. At certain times of the year, however, the water at the bottom is warmer than at the surface (winter), especially if there is ice cover, and no benefit would be realized.

There are also increased pumping and capital costs associated with pumping from the lake bottom in the form of longer pipes and increased friction losses through the pipes. So the answer to this question is unclear in general, but may be clarified for specific cases through appropriate calculations.

10. *Energy Conservation Solution 10.* [dks]. Based on the heating value of municipal solid waste given in the problem statement, the daily heating value of the processed solid waste is:

10. *Energy Conservation Solution 10.* [dks] (continued)

Q_r = (5,000 Btu/lb) (720 tons/d) (2,000 lb/ton) = 7.2 x 10^9 Btu/d

The energy value of the steam used for heating was given as 1,060 Btu/lb steam in the problem statement. Using this value, the mass of steam that can be produced from the energy produced from the burned solid waste is:

M_{steam} = (7.2 x 10^9 Btu/d)/(1,060 Btu/lb) = 6.8 x 10^6 lb steam/d

Any reduction in efficiency of conversion of energy from the solid waste due to heating up the water to make steam will reduce this calculated mass of steam. For example, typical conversion efficiencies for chemical potential energy (solid waste) to thermal energy (steam) is approximately 40%, resulting in a more realistic estimate of the mass of steam produced from the combustion of this 720 tons/d of municipal solid waste of:

M_{steam} actual = 0.40 (6.8 x 10^6 lb steam/d) = 2.7 x 10^6 lb steam/d

11. *Energy Conservation Solution 11.* [wma]. The following answer itemizes the positive and negative aspects of each energy generation method from the standpoint of their impact on the environment.

A contemporary <u>coal-fired steam boiler and electric generating facility</u> requires three primary raw materials, coal (the energy source), water (for steam, cooling, and probably emissions control), and lime or limestone (for emissions control of SO_2). Therefore, the potential impacts of raw material suppliers and waste management, as well as the potential impacts of coal combustion, must be considered. Some negative impacts may include:

- air pollution caused by SO_2, NO_x, particulate matter and CO_2 (global warming);
- water pollution from boiler operations (thermal pollution), surface or groundwater contamination from mining of coal and limestone; and
- land pollution from mining wastes and disposal of scrubber sludge, i.e., calcium sulfate;

Some positive impacts may include:

- producing huge amounts of electricity at one location where highly efficient environmental controls are cost effective;
- producing a potentially useful waste/by-product in the form of calcium sulfate; and

11. *Energy Conservation Solution 11*. [wma] (continued)

- producing potentially useful surplus heat, e.g., hot water, low pressure steam, etc.

The principal raw materials for a <u>nuclear power facility</u> are uranium and water (for cooling). The potential impacts of nuclear fission must be considered, as well as the potential impacts of uranium mining and processing. Some negative impacts may include:

- accidental release of radiation to the environment;
- thermal pollution of the cooling water supply;
- voluminous uranium mining and processing wastes, since only a very small percentage of uranium bearing ore is beneficially used; and
- difficult and costly storage and disposal of spent nuclear fuel, with a continuous, indefinite threat to the environment.

Some positive impacts may include:

- producing huge amounts of electricity at one facility, although highly toxic waste volumes are relatively small;
- virtually contaminant-free stack emissions if the plant is operating properly. No particulate emissions, heavy metals from fuel combustion, etc., are generated from nuclear power;
- no waste materials generated in the treatment of gas streams, so that the impact of nuclear power plants to the land are minimal when operated properly.

The principal materials used in <u>photovoltaic solar panel</u> power generation are the photovoltaic cells and water (for cooling). The potential impacts of solar cell power generation must be considered, as well as the potential impacts of the manufacturing of the cells themselves, and any batteries used to store energy during low solar radiation periods. Some negative impacts may include:

- generation of small amounts of hazardous waste in the manufacture of the solar cells;
- requirement for large land areas to provide adequate collector space to meet energy demands; and

11. *Energy Conservation Solution 11.* [wma] (continued)

- generation of hazardous waste from mining and production of materials in batteries used to store energy for later delivery to public when solar energy is not available.

Some positive impacts may include:

- producing small amounts of electricity at many separate facilities, and providing it to the power grid or storing it when it is not used by the generating unit;
- completely contaminant-free emissions during the energy generation process;
- no waste materials generated in the treatment of gas streams, water streams, etc.

The principal material required for a <u>hydroelectric power facility</u> is water (as energy supply and cooling). The potential impacts of hydroelectric power generation are primarily related to the impact of the physical structure, i.e., the dam, that is required for storage and management of the water used to power the turbines, and its effect on the local ecosystem in the river that has been converted to an electric power facility. Some negative impacts may include:

- disruption of the aquatic environment in the area where the facility is constructed;
- disruption of economy affected by disruption of ecosystem, and the environmental and socio/political implications of this;
- disruption of the aquatic environment upstream and downstream of the facility due to its impact on spawning, changing downstream water temperatures, etc.; and
- thermal pollution of the cooling water supply;

Some positive impacts may include:

- producing large amounts of electricity at one facility;
- virtually contaminant-free operations; and
- no waste materials produced in the generation of power with this type of system.

As indicated from the items listed above, there is currently no "perfect" energy source, and the choice of a particular source of

11. *Energy Conservation Solution 11*. [wma] (continued)

energy generation will be governed largely on the basis of social, political and cultural decisions, not on technical ones.

12. *Energy Conservation Solution 12*. [glh]. To solve this problem, first calculate the fuel usage each day in the U.S. from the data given in the problem statement. Since the number of miles traveled each day is taken as a constant, set it equal to $M_{trav/d}$.

$$M_{trav/d} = (gal/d) (mpg_{ave})$$
$$= (gal/d) (18 \text{ mpg})$$

Likewise, using the information that an increase in efficiency of 2 mpg would decrease the gallons of gasoline each day by 45 million gallons,

$$M_{trav/d} = (gal/d - 45 \times 10^6) (20 \text{ mpg})$$

Solving the two equations with two unknowns for gal/d yields a value of:

$$gal/day = 4.5 \times 10^8$$

On a per year basis, this equals:

$$gal/yr = (4.5 \times 10^8 \text{ gal/d}) (365 \text{ d/yr}) = 1.6 \times 10^{11} \text{ gal/yr}$$

To complete the problem, the 5% increase in efficiency must be considered. Logically, to travel the same number of miles/yr, only 95% of the original amount of fuel is used.

The savings due to MTBE use will be 5% of the fuel used times the cost of the fuel.

$$(0.05) (1.6 \times 10^{11} \text{ gal/yr}) (\$1.20/gal) = \$9.6 \times 10^9/yr$$
$$\text{(almost 10 billion dollars!)}$$

13. *Energy Conservation Solution 13*. [glh]. Refer to Problem 12 and its solution. While a savings of $10 billion/yr is a significant sum, and, with the positive effects on air pollution, it is almost certainly a valuable step, it is important to consider the cost of the MTBE used. This cost is the cost of MTBE per gallon (1 cent) times the total volume of gasoline used.

14. *Energy Conservation Solution 14*. [glh].

$$\text{total cost} = (cost/gal) (total \text{ gal used}) = (\$0.01) (1.6 \times 10^{11})$$
$$= \$1.6 \times 10^9/yr \text{ for MTBE additive.}$$

14. *Energy Conservation Solution 14.* [glh] (continued)

While this is a much more complex issue than can be fully treated here, one can still compare the fuel savings from use of this additive to the cost of its production. In dollars alone, the savings to the public is about 10 billion dollars annually ($9.6 x 10^9).

The cost is about one-sixth of this value ($1.6 x 10^9/yr), and so it would appear that the implementation of this energy-saving, pollution-reducing additive is fully justified. There are other issues, of course, many of which are difficult to assess. The oxygenation of the fuel improves the efficiency of combustion and the engine emits fewer hydrocarbons. How does this affect the CO_2 emissions? The SO_2 and NO_x emissions? Is there any danger to the public from the use of MTBE? There are recent reports of consumer health effects due to breathing the fumes of the substance. While these remain to be substantiated, this is a factor to be considered. Additionally, MTBE groundwater plumes of significant size have been observed below a number of leaking underground storage tanks due to MTBE's high solubility (40,000 mg/L!) and low affinity for aquifer solids. If MTBE is found to be a human health risk due to groundwater ingestion, significant liability for groundwater remediation and treatment will have been generated in its use as a fuel additive. Research being funded by the American Petroleum Institute, U.S. EPA, and others may answer these health risk questions.

Finally, are there any major effects to the national and world economies that need to be factored in? Are jobs lost in the petroleum industry? Are these counterbalanced by the jobs created in the manufacture of MTBE? The balance of trade should tip in favor of the United States, since less oil need be imported. Does this destabilize another nation or create a crisis elsewhere?

While the answer to these questions is beyond the scope of this workbook, it is important that the student (and the expert!) be alert to the broader picture. Due to our interlocking relationships with nature and other societies, changes in one part of our world may cause profound changes in another.

15. *Energy Conservation Solution 15.* (conservation, recycling, life-cycle analysis, LCA). [glh].

a. Obviously, with no recycling possible for Container A, this solution is the simplest. The cost for one million units of Container A is the cost for one unit times one million.

$$E_1 = (200 \text{ J/unit}) (10^6) = 2.0 \times 10^5 \text{ kJ}$$

b. Container B presents a somewhat different problem since it can be recycled up to five times.

15. *Energy Conservation Solution 15.* [glh] (continued)

However, in this problem, the consumer averages a return rate at each stage of the recycling effort of only 25%. The number of units of Container B, N, needed initially in order to have one million uses after a lifetime of five recycles at a recycling rate of 25% can be calculated as follows:

$$10^6 = N\,(1 + 0.25 + (0.25)^2 + (0.25)^3 + (0.25)^4 + (0.25)^5)$$

It is instructional to set up a table and examine the effect of the exponential in the equation.

Table 12
Number of Containers and Energy Requirements
for Container B

# Cycles	# Containers (N)	E_1 (10^5 kJ)	E_2 (10^5 kJ)	Total E (10^5 kJ)
1	800,000	3.60	0.50	4.1
2	761,905	3.43	0.60	4.0
3	752,941	3.39	0.62	4.0
4	750,733	3.38	0.63	4.0
5	750,183	3.38	0.61	4.0

Clearly the recycle rate is too low for the return on energy gained by recycling. Only a net savings on recycling of 2.5% is realized after one recycle. It is not worth the second recycling effort based upon an energy analysis.

c. Given what has been learned by solving the problem for Container B, it can be quickly seen what will occur in the application for Container C after only a few cycles. This is shown below in Table 13.

Table 13
Number of Containers and Energy Requirements
for Container C

# Cycles	# Containers (N)	E_1 (10^5 kJ)	E_2 (10^5 kJ)	Total E (10^5 kJ)
1	800,000	5.20	0.30	5.5
2	761,905	4.95	0.36	5.3
3	752,941	4.89	0.37	5.3

15. *Energy Conservation Solution 15.* [glh] (continued)

Given the rapidly diminishing returns, it is not necessary to carry out the calculations any further. The containers are "over-engineered" for this return rate. Although Container C can be recycled 20 times, energy-wise, it isn't worth it.

A comparison of the three containers shows that for one million container uses, the energy costs are:

Table 14
Comparison of Total Energy Requirements
for Containers A, B, and C

Containers	E_1 (10^5 kJ)	E_2 (10^5 kJ)	Total E (10^5 kJ)
A	2.0	0	2.0
B	3.4	0.6	4.0
C	4.9	0.4	5.3

The Vice President is advised to fire the engineers for bothering her with this presentation.

16. *Energy Conservation Solution 16.* [glh]. Clearly Container B will never be competitive with Container A. It takes more energy to recycle Container B than it does to manufacture Container A and throw it away.

For Container C, assume the best possible scenario in which the recycle rate is 100%. This will allow one to determine if there is any hope at all given the energy costs cited. If the recycle rate is 100%, then each Container A will have 20 uses. The total cost would be the initial cost plus the cost to remanufacture 19 times. There would be a huge initial savings because Quality Container Corporation would need only manufacture one-twentieth the number in order to get one million uses.

(1 x 10^6 total units)/20 = 5.0 x 10^4 units to be manufactured

The initial cost in terms of energy would be:

(5.0 x 10^4 units) (650 J/unit) = 3.25 x 10^4 kJ

The recycling costs would be:

(5.0 x 10^4 units) (19 cycles) (150 J/unit) = 1.42 x 10^5 kJ over the lifetime of the container

16. *Energy Conservation Solution 16.* [glh] (continued)

The total cost then would be:

$$3.25 \times 10^4 \text{ kJ} + 1.42 \times 10^5 \text{ kJ} = 1.75 \times 10^5 \text{ kJ}$$

This compares favorably with the value of 2.0×10^5 kJ for Container A. However, a 100% recycle rate is very unrealistic unless the incentives for the consumer are very high.

To determine the point at which a recycle rate (RR) for Container C will break even with the costs for Container A, the following expression is used knowing that the minimum energy cost for Container C to break even with Container A in this problem is 2.0×10^5 kJ

2.0×10^5 kJ = Units C (initially manufactured) (650 kJ/unit)
+ Units C $(RR^1 + RR^2 + RR^3 + ... + RR^{18} + RR^{19})$ (150 kJ/unit)

where RR = recycle rate (fraction of units recycled).

Setting up a table as before and with a recycle rate (RR) of 90% (RR = 0.90), we have:

Table 15
Required Number of Cycles and Energy Requirements
for Production of Container C to Break Even with that of Container A

# Cycles	# Containers (N)	E_1 (10^5 kJ)	E_2 (10^5 kJ)	Total E (10^5 kJ)
1	526,316	3.42	0.71	4.13
2	369,004	2.40	0.95	3.35
3	290,698	1.89	1.18	3.07
4	261,268	1.70	1.22	2.92
5	235,141	1.52	1.30	2.82
10	138,848	0.90	1.22	2.12
15	81,988	0.53	0.78	1.31
20	48,413	0.31	0.43	0.74

It appears that the number of cycles at a 90% recycle rate required to be competitive with Container A, for which the energy consumption is 2.0×10^5 kJ, is approximately 11 to 12. One would estimate that the recycle rate required to be competitive with the full 20 cycles that Container C is designed for would be about 80%. This is still a very high value and may be unrealistic. A better strategy would be to design a new container which is more competitive cost-wise and which is designed to withstand only 10 cycles.

Chapter 6

AIR QUALITY ISSUES

Suwanchai Nitisoravut

I. INTRODUCTION

The atmosphere has always been polluted to some extent through natural phenomena and/or human activities. In recent times, technological advancement, industrial expansion, and urbanization have been the major contributors to global air pollution. Air pollution adversely affects human health and property. A U.S. EPA (1994) report on national air quality and emission trends showed that emissions of nitrogen oxides (NO_x) increased 690%, emissions of volatile organic compounds (VOCs) increased by 260%, and sulfur dioxide emissions increased by 210% between 1900 and 1970. The Clean Air Act (CAA) has yielded substantial improvements in ambient air quality since its passage in 1970.

The CAA established two levels of National Ambient Air Quality Standards (NAAQSs), primary and secondary standards. Primary air quality standards are promulgated to protect public health, including the health of "sensitive" populations such as asthmatics, children, and the elderly. Secondary air quality standards are set to protect public welfare, including protection against decreased visibility, and damage to animals, crops, vegetation, and buildings.

The U.S. EPA has established NAAQS levels for six principal pollutants: carbon monoxide (CO), lead (Pb), nitrogen dioxide (NO_2), ozone (O_3), particulate matter whose aerodynamic diameter is less than or equal to 10 μm (PM-10), and sulfur dioxide (SO_2). PM-10 and SO_2 are regulated by primary standards for both short-term (24-hr or less) and long-term (annual average) averaging times. Short-term standards are established to protect the public from adverse health effects associated with acute exposure to air pollution, while long-term standards are established to protect the public from chronic exposures to pollution. Besides these six pollutants, "air toxic" compounds or hazardous air pollutants (HAPs) are also regulated following amendments made to the CAA in 1990. These HAPs are chemicals known to be acutely toxic, or that are known or suspected of causing cancer or other serious health effects (e.g., reproductive effects) or ecosystem damage. Examples of HAPs include dioxin, benzene, arsenic, beryllium, mercury, and vinyl chloride. The 1990 CAA Amendments list 189 compounds as HAPs.

The major sources of the six criteria pollutants for which NAAQSs exist are transportation fuel combustion, industrial processes, non-transportation fuel combustion, and natural sources such as wildfires. Unlike other criteria pollutants, ozone is not emitted directly to the air, but is created in the lower atmosphere when sunlight reacts with NO_x and VOCs. The air toxic compounds are emitted from a wide variety of sources including stationary and mobile point sources, and area sources.

While the quality of the ambient air has been improving, areas throughout the U.S. with significant air pollution are still widespread. Based on U.S. EPA monitoring data from 1994, approximately 62 million people in the U.S. reside in an area where at least one of the six criteria air pollutants exceed their NAAQS. Such a region is known in regulatory terms as a non-attainment area.

In addition to concerns regarding ambient air quality, indoor air quality has gained increasing public attention since there is now substantial evidence that many air pollutants tend to be at higher concentrations indoors than out. Findings from several studies suggest that indoor concentrations of NO_x, CO, and PM-10 can routinely exceed NAAQS levels in many non-occupational settings including residential, commercial, institutional, and transportation-related microenvironments.

Indoor air pollution sources that release gases or particulate matter into the air are the primary cause of indoor air quality problems. These indoor air pollution sources include: combustion sources (i.e., oil, gas, kerosene, wood, or coal heating and cooking systems, and tobacco products); building materials and furnishings (i.e., deteriorated asbestos-containing insulation, new carpet, cabinetry or furniture made from pressed wood products); household cleaners and personal care or hobby-related chemicals; central heating and cooling systems and humidifiers; and outdoor sources of ambient air pollution that can result in concentrated indoor air levels due to building ventilation systems. Common indoor air pollutants include: radon, environmental tobacco smoke (secondhand smoke), microbiological contaminants, CO, NO_2, organic gases (e.g., formaldehyde, pesticides, other solvents, cleaners, and disinfectants), respirable particles, asbestos, and lead. Unlike with ambient air quality, there are no indoor air quality standards for non-industrial workplace environments as it is currently felt that enforcing strict indoor air quality standards would be impractical.

There are three basic strategies for both ambient and indoor air quality control: prevention, dispersion or ventilation, and collection and treatment. The formation of pollutants in combustion processes is sometimes controllable through one or more of the following approaches: replacing or altering the fuel used in the process, changing the production process or equipment, and improving process operation and maintenance practices. Often, however, generation of pollutants is unavoidable. If generated, it may be possible to dilute the concentration of a pollutant to acceptable levels through atmospheric dispersion (in ambient air quality situations) or dilution ventilation (in indoor air quality situations). While dilution alone is not generally an acceptable air quality control technique for ambient air quality management, the magnitude of pollutant dilution is accounted for when determining the treatment performance required for a control device used on a stack discharging into the ambient atmosphere.

If collection and treatment of air pollutants are required, a variety of air quality control equipment is available for the removal of specific types of contaminants. Some equipment is designed specifically to remove particulates from the gas stream (i.e., gravity settlers, cyclones, scrubbers, baghouses, and electrostatic precipitators), while other equipment is designed for the elimination of gases and odors (i.e., adsorbers, absorbers, condensers, and oxidation units).

This chapter provides examples of both ambient and indoor air quality problems that include issues related to pollutant dispersion and transport, exposure assessment, and design and analysis of a variety of air pollution control equipment.

REFERENCES

The reader is referred to the following text/reference books for additional information on air quality issues.

1. **Holmes, G., Singh, B. and Theodore, L.**, Environmental risk assessment, in *Handbook of Environmental Management and Technology*, Wiley-Interscience, New York, 1993, chap. 33.
2. **U.S. EPA**, The Inside Story - A Guide to Indoor Air Quality, EPA/400/1-88/004, U.S. Environmental Protection Agency, Washington, D.C., 1988.
3. **U.S. EPA**, National Air Quality and Emissions Trends Report, EPA/454/R-95/014, U.S. Environmental Protection Agency, Washington, D.C., 1988.

II. PROBLEMS

1. *Air Quality Issues Problem 1*. (indoor air quality). [pg]. List the factors that affect the level or concentration of a pollutant in an indoor air environment.

2. *Air Quality Issues Problem 2*. (indoor air quality, ambient air quality, health effects, costs, recent developments). [pg, hb]. Indoor air quality is a relatively recent concern in environmental management.

    a.   Explain why indoor air quality is a concern.
    b.   Explain what has caused indoor air quality to be of greater concern now than in the past.
    c.   Describe some of the immediate and long-term health effects of indoor air quality exposure.
    d.   What are some of the costs of indoor air quality problems?
    e.   Compare indoor air pollution with ambient air pollution. Show why indoor air quality can be of greater concern than ambient air quality.

3. *Air Quality Issues Problem 3*. (indoor air quality, air exchange rates). [hb]. The levels of contaminants in indoor air are a result of the combined influences of conditions which produce, mix, or remove those contaminants. These conditions are often classified into five major categories: sources of contaminants, air exchange rates, contaminant removal mechanisms, volume of the structure, and the mixing efficiency of the indoor area.

    a.   Discuss in some detail the term "air exchange rates." Your answer should explain what is meant by the term "air exchange rate" and then discuss some of the variables that affect it. The discussion should address the following terms: infiltration, exfiltration, wind effects, stack effects, combustion effects, natural ventilation, and forced ventilation.
    b.   Discuss in some detail the sources of contaminants in indoor air. Your answer should include indoor sources, outdoor sources, structural sources, product sources, and sources due to activities carried out in the indoor space.

4. *Air Quality Issues Problem 4*. (indoor air quality). [pg]. Explain briefly the status of regulations and standards that have been established for the residential indoor air environment where pollutant levels may be higher than those found in ambient air.

5. *Air Quality Issues Problem 5.* (indoor air quality, formaldehyde, chemical properties, physical properties, health effects, carcinogenic, sources). [hb]. Formaldehyde is a chemical which is produced in great quantities and is very widely used in many materials and products that could affect indoor air quality. It is also a chemical whose properties make it of considerable concern in any environmental management system that addresses indoor air quality.

    a.   List some physical and chemical properties of formaldehyde.
    b.   Discuss some of the health concerns associated with formaldehyde.
    c.   List as many potential categories and specific sources of formaldehyde as you can that could have an impact on indoor air quality.

6. *Air Quality Issues Problem 6.* (indoor air quality, radon, asbestos, lead, VOCs). [pcy]. Identify the major contaminants affecting indoor air quality and provide a potential source of each contaminant.

7. *Air Quality Issues Problem 7.* (asbestos, U.S. EPA, regulation). [mh]. Prolonged exposure to asbestos is believed to cause a number of ailments including asbestosis, lung cancer, and mesothelioma. In order to solve this problem, the U.S. government has set up five different regulatory bodies to monitor the asbestos threat. List the five regulatory bodies and their primary responsibilities relative to the asbestos problem.

8. *Air Quality Issues Problem 8.* (indoor air quality). [pg]. Radon is a cancer-causing, radioactive gas that causes many thousands of deaths each year in the U.S.

    a.   Explain how radon is formed, how it disperses within the environment, and how it can make its way into an indoor atmosphere.
    b.   Briefly explain the main human exposure pathways and human health effects that can result from radon exposure.

9. *Air Quality Issues Problem 9.* (indoor air quality, radon, house inspection). [sl]. An inspector was asked to check an old house for seal openings that could permit radon entry. What areas should be checked by the inspector?

10. *Air Quality Issues Problem 10.* (indoor air quality, humidity, temperature). [pcy]. What range of indoor temperature and humidity are comfortable for most people?

11. *Air Quality Issues Problem 11.* (indoor air quality, dilution ventilation). [sn]. Cell immobilization technology has been used widely in water and wastewater treatment applications, particularly in laboratory-scale studies. A number of immobilization carriers have been employed, including calcium alginate and cellulose-triacetate. Toluene is required as a hardening agent in the preparation of the cellulose-triacetate carrier.

Assume that a student has prepared a cellulose-triacetate immobilization carrier and has kept it in four trays (8 in wide by 15 in long by 2 in deep) filled with toluene. The trays have been left on the table top in the media prep room. It was observed that toluene in each tray had decreased by 10% (v/v) after 1 hour.

    a.   Estimate the maximum concentration of toluene in the room in units of parts per million by volume (ppmv) and mg/m³.

    b.   Assuming that air distribution through the zone of contamination is relatively poor (safety factor, K, of 10 should be used) and the acceptable toluene indoor air concentration is its threshold limit value (TLV) of 50 ppmv, what is the minimum volume of air required to dilute the evaporated vapor volume to reach a safe indoor air concentration?

    c.   Assume that the student turns on the hood with the door fully open. The average face velocity of the hood is 100 fpm. Will this dilution ventilation adequately reduce the toluene level in the room air to or below its TLV?

The following data are provided to solve the problem:
- Molecular Weight (MW) of toluene = 92.13 g/gmol
- Specific gravity of toluene = 0.866
- Vapor pressure of toluene at 20°C = 18.4 mm Hg
- Room temperature = 20°C
- Room dimensions = 20 ft by 50 ft by 12 ft
- Hood dimensions = 2 ft high by 4 ft long by 2 ft deep

Also, the following equation can be used to estimate the required dilution air flow rate:

$$Q = \frac{(F)\,(\text{specific gravity})\,(W)\,(K)\,(10^6)}{(MW)\,(TLV)}$$

11. *Air Quality Issues Problem 11*. [sn] (continued)

where Q = dilution air flow rate, cfm; W = amount of liquid used per time interval, gal/min; F = conversion factor = 3222 for W in units of gal/min; TLV = acceptable air concentration, ppmv; and K = safety factor.

12. *Air Quality Issues Problem 12.* (exposure assessment, water-air equilibrium, indoor air quality). [sl]. It has been long known that people are more likely to be exposed to VOCs indoors than outdoors. For example, levels of certain VOCs present in indoor air have been found to be 10 times higher than in outdoor air. However, daily indoor inhalation of VOCs may not result exclusively from the indoor air itself. Certain routine activities such as taking a shower or using a dishwasher could bring additional amounts of VOCs into the indoor air if the tap water is contaminated. Note that besides inhalation, VOCs can enter a human body via ingestion and absorption through skin.

It is interesting to calculate the mass of VOC that enters indoor air from an "innocent" shower stall. The equilibrium between water and air may be described by Henry's law:

$$y = m\,C$$

where y = the equilibrium gas phase contaminant concentration, mg/L; C = the aqueous phase contaminant concentration, mg/L; and m = the Henry's constant with a value of 0.3 for this problem.

An idealized shower stall is assumed to have steady-state volumetric flow rates of water (Q_L = 12 L/min) and air (Q_G = 35 L/min). The shower air volume (V_G = 1500 L) is assumed to be thoroughly mixed. At t = 0, air entering the shower has a constant contaminant concentration y_{in} = 0.015 mg/L and the water entering the shower has a constant contaminant concentration C_{in} = 0.4 mg/L. Estimate the shower air contaminant concentration, y, after a 15 minute shower.

13. *Air Quality Issues Problem 13.* (dispersion, permit, source review). [dj]. You are working for the U.S. EPA Region IX office in San Francisco and are checking the plans for a proposed new hazardous waste incinerator to be located 50 miles northeast of Las Vegas. The plans show that the physical height of the stack is twice the height of the nearest surrounding buildings and obstacles. The incinerator plans also show that the designers want to double the flow rate of stack gases by injecting dilution air. They claim that, regardless of atmospheric conditions, injection of dilution air will decrease ground-level concentrations by one-half.

Are these good design decisions? You may consult any of several references on stack design in formulating your answer.

13. *Air Quality Issues Problem 13*. [dj] (continued)

Possible references include:

> **Cooper, C.D. and Alley, F.C.**, Dispersion modeling, in *Air Pollution Control: A Design Approach*, 2nd edition. Waveland Press, Prospect Heights, Illinois, 1994, chap. 19.
> **Holmes, G., Singh, B. and Theodore, L.**, Atmospheric dispersion, in *Handbook of Environmental Management and Technology*, Wiley-Interscience, New York, 1993, chap. 7.
> **Theodore, M.K. and Theodore, L.**, Atmospheric dispersion modeling, in *Major Environmental Issues Facing the 21st Century*, Prentice-Hall, New York, 1996, chap. 10.

14. *Air Quality Issues Problem 14*. (electrostatic precipitator, collection efficiency, particulates). [jr]. The Deutsch-Anderson equation, commonly employed in the design of electrostatic precipitators (ESPs), is given by:

$$\eta = 1 - \exp\left(-\frac{w\,A}{q}\right)$$

where η = fractional collection efficiency; w = effective drift velocity, length/time; A = collection area, length^2; and q = volumetric flow rate, length^3/time.

The collection efficiency is defined as:

$$\eta = \frac{\text{mass of particulate matter collected (rate basis)}}{\text{mass of particulate matter in the inlet stream (rate basis)}}$$

An electrostatic precipitator (ESP) with 60,000 ft^2 of collection area is cleaning 5,000 actual cubic feet/min (acfm) of a particulate-laden gas stream. The initial particulate loading has been measured at 4.41 grains/actual cubic foot (gr/acf), while an outlet loading of 0.04 gr/acf must be achieved in order to comply with state regulations.

- a. What is the minimum collection efficiency that will satisfy the regulations?
- b. What is the minimum average drift velocity required to achieve the collection efficiency calculated in Part a?

15. *Air Quality Issues Problem 15*. (air quality, benzene, anthracene, desorption, volatilization, mass transfer). [dks]. Benzene (C_6H_6) and anthracene ($C_{14}H_{10}$) desorption were studied simultaneously in a pilot-scale surface-agitated vessel. Experimental conditions were: 0.075 HP motor, single-blade 3 in. dia.

15. *Air Quality Issues Problem 15.* [dks] (continued)

impeller, 50 in. dia. vessel, water depth = 8.75 in., floor fan to simulate wind at 3 to 4 mph, temperature 18 to 24.5°C, and impeller speed = 465 to 545 rpm. Experimental results are shown in Table 16.

Table 16
Simultaneous Absorption and Desorption Data
for Benzene and Anthracene

Benzene desorption data		Anthracene desorption data	
Time (min)	Concentration (mg/L)	Time (min)	Concentration (mg/L)
0	1,585	0	75
18	1,570	1.5	70
35	1,460	3.87	65
52	1,240	5.83	57
67	1,085	8.13	52
90	1,025	10.9	46
127	890	14.4	40
157	745	20.1	35
363	330	26.6	27

Note: Use a first-order model for benzene and anthracene stripping as follows:

$$C = C_o e^{-Kla\, t}$$

a. Determine the experimental desorption coefficient, Kla, for benzene (hr^{-1}).
b. Determine the experimental absorption coefficient, Kla, for anthracene (hr^{-1}).
c. Explain the difference, if any, in the coefficients for desorption and absorption.
d. Discuss any implications this might have for surface impoundments of polynuclear aromatic hydrocarbons (anthracene is a polynuclear aromatic hydrocarbon).

16. *Air Quality Issues Problem 16.* (particulate control devices, gaseous control devices, cyclone, fabric filters, electrostatic precipitator). [kg]. What are the common control devices used to remove particulate matter and gaseous pollutants from contaminated gas streams?

17. *Air Quality Issues Problem 17.* (particulate control devices). [kg]. Explain the advantages and disadvantages of the following particulate control devices:

17. *Air Quality Issues Problem 17.* [kg] (continued)

      a.   electrostatic precipitators.
      b.   fabric filters.
      c.   wet scrubbers.

18. *Air Quality Issues Problem 18.* (NSPS, emissions). [sn]. A 500 MW steam electric power plant burns coal that contains 77.2 wt% C, 5.2 wt% H, 1.2 wt% N, 2.6 wt% S, 5.9 wt% O, and 7.9 wt% ash with a heating value of 30,000 kJ/kg. Calculate the daily emission of particles, SO_2, NO_x, and CO_2 permitted by the New Source Performance Standards (NSPS). The thermal efficiency of the plant is 40%.

The NSPS permitted emission rates for coal in electric steam power plants are as follow:

      Particulate matter: 13 g/106 kJ
      NO_x: 260 g/106 kJ
      SO_2: 260 g/106 kJ (dependent upon %S)

19. *Air Quality Issues Problem 19.* (stack height, plume rise). [pcy]. The Grand Gulf power plant with a 90 m tall, 4 m dia. stack emits gas at 95°C with an exit velocity of 16 m/s. What is the plume rise with a wind speed of 5 m/s, using the Bryant-Davison stack rise equation? Assume the air temperature is 25°C.
    The Bryant-Davison equation is:

$$dH = S \, (V_s/U)^{1.4} \, (1 + dT/T_s)$$

where dH = the rise of the plume above the stack exit, m; S = the inside stack diameter, m; V_s = stack gas velocity, m/s; U = wind speed, m/s; dT= the stack gas temperature - the ambient air temperature, K; and T_s = the stack gas temperature, K.

20. *Air Quality Issues Problem 20.* (stack height, plume rise). [pcy]. Use the conditions provided in Problem 19 and an atmospheric pressure of 1,050 mb. What is the effective stack height calculated from the Holland equation for neutral stability?
    The Holland equation for stack plume rise is:

$$dH = \left(\frac{V_s S}{U} \right) \left[1.5 + 2.68 \times 10^{-3} \, P \left(\frac{dT}{T_s} \right) S \right]$$

where P = atmospheric pressure, mb; and all other terms are as defined in Problem Statement 19.

21. *Air Quality Issues Problem 21.* (baghouse, particulate control device). [sxl]. An 850 MW coal-fired power plant produces 1.7 million cfm of stack gases. Find:

    a. How many square feet of filter surface area would be required for a baghouse to remove the particulates from this flow?

    b. How many bags are required in this baghouse?

Typical data for a cylindrical baghouse filter are given below.

    Length = 38 ft
    Diameter = 1 ft
    P= 675 N/m^2
    V = 0.02 m/s
    Cake density = 1,000 kg/m^3
    Cake thickness = 0.15 mm
    k = 0.71 x 10^{-13} m^2
    α = 5 x 10^{-7} m^2

22. *Air Quality Issues Problem 22.* (baghouse, particulate control device). [sxl]. Refer to the solution to Problem 21. If each bag must be cleaned once an hour and the cleaning time is 3 minutes how many bags must be used now to ensure the surface area requirements are met?

23. *Air Quality Issues Problem 23.* (data analysis, log normal distributions). [fwk]. During an 8-year period in the Washington, D.C. area the ambient air SO_2 concentrations were found to vary as shown in Table 17.

Table 17
Historical SO$_2$ Data from Washington, D.C.

1-Hour Average SO$_2$ Concentration (ppmv)	% Time Concentration Occurred
0.10 to 0.45	10.0
0.07 to 0.10	10.0
0.06 to 0.07	10.0
0.04 to 0.06	10.0
0.03 to 0.04	10.0
0.02 to 0.03	20.0
0.01 to 0.02	20.0
0.00 to 0.01	10.0

It has been decided that public warnings are necessary each time the concentration exceeds 0.25 ppmv. How often will this occur based on these historical data?

24. *Air Quality Issues Problem 24.* (effective stack height, plume rise, general public). [dj]. A consulting engineering company has prepared a design analysis of a proposed stack for a new gas-fired electrical generator at the power plant that employs you as a public information officer. The consultant has concluded that, given the worst-case atmospheric conditions expected at your facility, the worst-case effective stack height will be 140 m, and that, at this stack height, the probability of exceeding the National Ambient Air Quality Standard is 1.5 x 10^{-6}. The consultant used the ISC (Industrial Source Complex) model to carry out the air quality calculations for the permit application for the new generator.

The state air quality control board (AQCB) holds a public meeting to review and invite comment on the application. The AQCB is comprised of elected public officials, none of whom have technical backgrounds. Past relations between the utility and the AQCB have been contentious, and the utility is anxious to improve this relationship. You were hired as part of that plan to improve the relationship.

A public comment period is held after presentations by the utility, the consultant, and AQCB staff (who are favorable to the application, as they believe that all the technical work in the application is competent). During the public comment period, a concerned citizen goes to the floor microphone and states that "the application shows that the stack is 110 m tall, but that the effective stack height is 140 m." The citizen is concerned that the utility "may not have its act together, as it can't even get the stack height right." The chair of the AQCB turns to you and asks you to reply to the citizen's concern. Your consultant has left the meeting to catch a plane home, so it is up to you to respond verbally to the citizen's concerns.

What do you say? Write two or three paragraphs that serve as your verbal response to the citizen.

Remember that, above all, you don't want anything to happen that will delay approval of the application. Also, you don't want tomorrow's press reports to show that your company ignored or disregarded the citizen's concerns. You will have to explain technical concepts in layman's language, and above all, be pleasant, constructive and sympathetic.

If you do not have an air quality background, two books containing non-technical information about air pollution stack design may be helpful in formulating your answer: These are:

Holmes, G., Singh, B. R., and Theodore, L., Atmospheric dispersion, in *Handbook of Environmental Management and Technology*, J. Wiley and Sons, New York, 1993, chap. 7.

Theodore, M.K. and Theodore, L., Atmospheric dispersion modeling, in *Major Environmental Issues Facing the 21st Century*, Prentice-Hall, New York, 1996, chap. 10.

III. SOLUTIONS

1. *Air Quality Issues Solution 1.* [pg]. Many factors such as source strength, indoor volume, occupant's behavior, ventilation rates, and thermal efficiency factors affect the level and composition of pollutants found in an indoor atmosphere.

2. *Air Quality Issues Solution 2.* [pg, hb].

   a. Indoor air quality (IAQ) is a major concern because indoor air pollution may present a greater risk of illness than exposure to outdoor pollutants. People spend 75 to 90% of their time indoors. This situation is compounded as sensitive populations, e.g., the very young, the very old, and sick people who are potentially more vulnerable to disease, spend many more hours indoors than the average population.

   b. Indoor air quality problems have become more serious and of greater concern now than in the past because of a number of developments that are believed to have resulted in increased levels of harmful chemicals in indoor air. Some of those developments are the construction of more tightly sealed buildings to save on energy cost, the reduction of the ventilation rate standards to save still more energy, the increased use of synthetic building materials and synthetics in furniture and carpeting that out-gas harmful chemicals, and the widespread use of new chemically formulated personal care products, pesticides, paints and cleaners.

   c. Some of the immediate health effects of indoor air quality problems are irritation of the eyes, nose and throat, headaches, dizziness and fatigue, asthma, pneumonitis, and "humidifier fever." Some of the long-term health effects of indoor air quality problems are respiratory diseases and cancer. These are most often associated with radon, asbestos, and second hand tobacco smoke.

   d. The U.S. EPA, in a report to Congress in 1989, estimated that the costs of indoor air quality problems were in the tens of billions of dollars per year. The major types of costs from indoor air quality problems are direct medical costs, lost productivity due to absence from the job because of illness, decreased efficiency on the job, and damage to materials and equipment.

2. *Air Quality Issues Solution 2.* [pg, hb] (continued)

    e. Outdoor air pollution and indoor air pollution share many of the same pollutants, concerns and problems. Both can have serious negative impacts on the health of the population. Not too many years ago, it was a common practice to advise people with respiratory problems to stay indoors on days when pollution outdoors was particularly bad. The assumption was that the indoor environment provided protection against outdoor pollutants. Recent studies conducted by the U.S. EPA have found, however, that the indoor levels of many pollutants are often two to five times, and occasionally more than 100 times, higher than corresponding outdoor levels. Such high levels of pollutants indoors is of even greater concern than outdoors because most people spend more time indoors than out. Some estimates indicate that most people spend as much as 90% of their time indoors. Indoors is where most people work, attend school, eat, sleep, and even where much recreational activity takes place.

3. *Air Quality Issues Solution 3.* [hb].

    a. The "air exchange rate" is the rate at which indoor air is replaced with outdoor air. The units of the air exchange rate are "air changes per hour" or "ach." If the volume of air in a building is replaced twice in 1 hour, the air exchange rate would be two. If the volume of air in a building is replaced once in 2 hours, the air exchange rate would be 0.5. The air exchange rate can be calculated by dividing the rate at which outdoor air enters the building in m^3/hr (or ft^3/hr) by the volume of the building in m^3 (or ft^3). If the air exchange rate were 1 ach, it would not mean that every molecule of indoor air would have been replaced at the end of 1 hour. Just which molecules were replaced would depend on a number of factors. Some of those factors are infiltration, exfiltration, wind effects, stack effects, combustion effects, natural ventilation, and forced ventilation.
Infiltration and exfiltration refer to the uncontrolled leakage of air into or out of the

3. *Air Quality Issues Solution 3*. [hb] (continued)

building through cracks and other unintended openings in the outer shell of the building. In addition to leakage around windows and doors, infiltration and exfiltration can occur at points such as openings for pipes, wires and ducts. The rate of infiltration and exfiltration can vary greatly depending on such factors as wind, temperature differences between indoors and outdoors, as well as the operation of chimneys and exhaust fans.

Wind effects result from wind striking one side of a building causing positive pressure on that side and lower pressure on the opposite side (the leeward side). Air is forced into the building on the windward side and out the leeward side. Some buildings may be somewhat protected from wind effects by terrain, trees, and other buildings.

The tendency of warm air to rise in a room or through the levels of a multilevel building results in what is known as stack effects.

In winter, when there is a large temperature difference between indoor and outdoor air, rising warm air escapes through openings at the top of the building and outdoor air is drawn in at the bottom of the building. The effect is usually less pronounced in the summer because of smaller temperature differences and the direction of the flow may be reversed.

Combustion effects often arise from fires in fireplaces, stoves, and heating systems. The combustion uses up indoor air which causes pressure in the building to drop. Outdoor air is then drawn in. This effect can double infiltration rates. Use of outdoor air in a heating system or fireplace substantially reduces this effect.

Natural ventilation is air that is drawn into a building through windows, doors and other controlled openings. Natural ventilation results from wind striking the building and/or temperature differences between the outdoor air and the indoor air.

Forced ventilation refers to drawing air into a building through fans and ducts. The effectiveness in removing contaminants from indoor air

3. *Air Quality Issues Solution 3.* [hb] (continued)

      by the use of forced ventilation can vary widely. Fans used to exhaust specific sources of pollutants (such as the kitchen stove) can be very effective. Most forced ventilation systems are used to circulate air conditioned air. Whole house fans and the forced ducted ventilation systems of large buildings must be carefully balanced by air supply to prevent backdrafts of contaminants from chimneys and heating plants.

   b. Indoor sources of indoor air contaminants include consumer and commercial products, building or structural materials, and personal activities.

      Among the consumer and commercial products that release pollutants into the indoor air are pesticides, adhesives, cosmetics, cleaners, waxes, paints, automotive products, paper products, printed materials, air fresheners, dry cleaned fabrics, and furniture. In addition to the "active" ingredient in all the products mentioned, many products contain so-called inert ingredients that are also contaminants when released into indoor air. Examples would be solvents, propellants, dyes, curing agents, flame retardants, mineral spirits, plasticizers, perfumes, hardeners, resins, binders, stabilizers, and preservatives. Aerosol products produce droplets which remain in the air long enough to be inhaled. This allows the inhalation of some chemicals which would not be volatile enough to be inhaled otherwise. One consumer product that produces indoor air contaminants and merits special mention is tobacco smoke.

      The single most important building or structural source of contaminants is formaldehyde from building materials such as plywood, adhesives, insulating materials such as urea formaldehyde foam, floor and wall coverings. Depending on the individual type of construction and mainten-ance practices, there can be many other building sources of indoor air quality problems. Damp or wet wood, insulation, walls and ceilings can be breeding places for allergens and pathogens that can

3. *Air Quality Issues Solution 3.* [hb] (continued)

become airborne. Allergens and pathogens can also originate from poorly maintained humidifiers, dehumidifiers, and air conditioners. If the building has openings to the soil, radon can enter the building in those areas where radon occurs. The building's heating plant can be the source of contaminants such as CO and NO_x. Automobile exhaust from attached garages is another source of carbon monoxide and nitrogen oxides. Particulates such as asbestos from crumbling insulation and lead from the sanding of lead based paints are additional contaminants that can become part of the indoor air.

Personal activities can be sources of indoor air contaminants such as pathogenic viruses and bacteria, and a number of harmful chemicals such as products of human and animal (pet) respiration. Houseplants can release allergenic spores. Pet products, such as flea powder, can be sources of pesticides and pets produce allergenic dander when they lick themselves.

Outdoor sources of indoor air contaminants are widely varied. Polluted outdoor air can enter the indoor space through open windows, doors and ventilation intakes. Most outdoor air is less contaminated than indoor air. In some cases, however, such as with a nearby smokestack, a parking lot, heavy street traffic, or an underground garage, outdoor air can be a significant source of indoor contaminants. Outdoor pesticide applications, barbecue grills, and garbage storage areas can also bring outdoor contaminants into the building if placed close to a window or door, or the intake of a ventilation system. Improper placement of the intake of a ventilation system near a loading dock, parking lots, the exhaust from restrooms, laboratories, manufacturing spaces, and other exhausts of contaminated air is a major source of indoor air pollution. Other outdoor sources of indoor air pollutants are hazardous chemicals entering the structure from the soil. Examples are radon gas, methane and other gases from sanitary landfills, and vapors from leaking underground storage tanks of gasoline, oil, and other

3. *Air Quality Issues Solution 3.* [hb] (continued)

>   chemicals penetrating into basements. Polluted water can give off substantial quantities of harmful chemicals during showering, dish-washing and similar activities.

4. *Air Quality Issues Solution 4.* [pg]. The U.S. EPA is responsible for setting and enforcing the NAAQSs that are designed to protect the health of the general public with an adequate margin of safety. The Occupational Safety and Health Administration adopts and enforces standards for industrial work environments which are designed such that no employee will suffer material impairment of health or functional capacity. However, federal responsibility and authority for indoor air quality (IAQ) in the non-workplace are less defined. At present there are no federal IAQ standards for the general public. The General Service Administration (GSA) has established "GSA Action Levels" for indoor air contaminants. The American Society of Heating Refrigeration and Air Conditioning (ASHRAE) recommends one tenth of the occupational standards to be used as the safe level for chemicals in the indoor atmosphere.

5. *Air Quality Issues Solution 5.* [hb].

>   a. <u>Physical/chemical properties</u> - Formaldehyde is a flammable colorless gas with a characteristic odor. Most people can detect the odor at concentrations as low as 1 ppmv. Some people can detect the odor at concentrations as low as 0.05 ppmv. The odor is unpleasant and tends to produce a choking response. It readily evap-orates from its mixtures.
>
>   b. <u>Health Concerns</u> - Formaldehyde is a known irritant and sensitizer. Inhalation symptoms include burning and stinging in the nose and throat as well as pain in the eyes. Coughing, chest tightness and wheezing can occur at higher concentrations. Skin contact can produce irritation, dermatitis, and sensitization. Other effects include nosebleeds, runny noses, persis-tent swelling of nasal turbinates, headaches, fatigue, memory and concentration problems, nausea, dizziness, and breathlessness.
>   Formaldehyde has also been reported to be associated with altered reproductive function in women and fetotoxic effects, but these effects have not as yet been confirmed. It has also been

5. *Air Quality Issues Solution 5.* [hb] (continued)

shown to be mutagenic in some test systems, including humans. Animal tests and human epidemiological studies suggest that formaldehyde should be presumed to pose a carcinogenic risk to humans. The Federal Panel on Formaldehyde (1982), the International Agency for Research on Cancer (1982), the Consumer Product Safety Commission (1986), and the U.S. EPA (1987) have concluded that formaldehyde does indeed pose such a threat. In 1987, the U.S. EPA classified formaldehyde as a "Probable Human Carcinogen" (Group B1) based on sufficient animal and limited human evidence and other supporting data. Additional studies are being conducted and the issue is by no means settled.

c. The following are some categories of potential sources and specific potential sources of formaldehyde that could affect indoor air quality.

- <u>Pressed Wood Products</u> - Hardwood, plywood, particle board, medium density fiberboard (MDF), and decorative paneling.
- <u>Insulation</u> - Urea-formaldehyde foam insulation (UFFI) and fiberglass made with formaldehyde binders.
- <u>Combustion Sources</u> - Natural gas, kerosene, tobacco, and automobile exhaust.
- <u>Paper Products</u> - Grocery bags, waxed paper, facial tissues, paper towels, and disposable sanitary products.
- <u>Stiffeners, Wrinkle Resisters, and Water Repellents</u> - Floor coverings (rugs, linoleum, varnishes, plastics), carpet backings, adhesive binders, fire retardants, and permanent press textiles.
- <u>Miscellaneous</u> - Plastics, cosmetics, deodorants, shampoos, disinfectants, starch-based glues, adhesives, laminates, paints, fabric dyes, inks, fertilizers, and fungicides.

6. *Air Quality Issues Solution 6.* [pcy]. Some major indoor air pollutants and their corresponding sources include:

- VOCs - cleaning solvents
- Pesticides - disinfectants

6. *Air Quality Issues Solution 6*. [pcy] (continued)

- Lead - lead based paint
- Asbestos - ceiling and floor tiles
- Radon - water
- Biological contaminants - plants
- Tobacco smoke - tobacco products
- Particulates (respirable) - fireplace
- Polycyclic aromatic hydrocarbon contaminants - wood stoves
- SO_2 - kerosene heaters
- Formaldehyde - urea formaldehyde foam insulation
- CO - vehicle exhaust
- CO_2 - improper operating gas ranges and heaters
- NO_x - vehicle exhaust

7. *Air Quality Issues Solution 7*. [mh]. The five regulatory bodies and their primary responsibilities related to the asbestos problem in the U.S. are listed below.

1. Occupational Safety and Health Administration (OSHA) sets limits for worker exposure to asbestos on the job.
2. The Food and Drug Administration (FDA) is responsible for preventing asbestos contamination in foods, drugs, and cosmetics.
3. The Consumer Product Safety Commission (CPSC) regulates asbestos in consumer products. It has already banned the use of asbestos in drywall patching compounds, ceramic logs, and clothing. The CPSC is now studying the extent of asbestos use in consumer products in general, and is considering a ban on all nonessential product uses that can result in the release of asbestos fibers.
4. The Mine and Safety and Health Administration (MSHA) regulates mining and milling of asbestos.
5. The U.S. Environmental Protection Agency (U.S. EPA) regulates the use and disposal of toxic substances in air, water, and land, and has banned all uses of sprayed asbestos materials. The effect of cumulative exposure to asbestos has been established by dozens of epidemiological studies. In addition, the U.S. EPA has issued standards for handling and disposing of asbestos containing materials.

8. *Air Quality Issues Solution 8*. [pg].

a. Radon is not a product of modern technology but a naturally occurring compound that poses lifetime lung cancer risk greater than most commercial carcinogenic chemicals. Radon is

8. *Air Quality Issues Solution 8.* [pg] (continued)

naturally emitted because of the abundant occurrence of radium 226 in the earth's crust by radioactive decay processes. Radon is commonly found in uranium ores, phosphate rock and a number of common minerals such as granite, schist, gneiss, and even limestones. Radon is produced when uranium molecules release protons and neutrons from their nuclei. In the process the uranium molecule changes first into thorium, then radium, and then radon. Radon is an inert gas with a half-life of 3.8 days.

Radon is a noble gas and thus freely moves in the enclosed indoor atmosphere by diffusion as well as convection. Indoor air currents, cracks in slabs, basement floors and walls, water sumps and building pressure conditions are the mechanisms that transfer and disperse radon in the indoor atmosphere. Radon has been estimated to account for 5,000 to 20,000 cases of lung cancer in the U.S. alone.

b. Occupational health effects related to radon have been well documented. Radon is an inert gas with a half-life of 3.8 days and is not hazardous by itself. It is extremely unreactive, but when it undergoes radioactive decay, a series of short-lived radon progeny are produced. These radon daughter products create the greatest health hazards. They are extremely small solid particulates and are electrically charged. These radon progeny can very easily attach themselves to large airborne particles including dust, shed skin and cigarette smoke particles, and enter the respiratory system and deposit in the lungs. Radon has special affinity for smoke particles and produces synergistic effects. Inhalation is the route of entry for radon. When deposited in the non-ciliated portion of the lungs, these particles damage the surrounding tissue and cause lung cancer. High exposure levels of radon are associated with incidences of lung cancer that exceed cancer due to all industrial chemicals.

9. *Air Quality Issues Solution 9.* [sl]. The inspector should at least check the following areas for radon entry points.

9. *Air Quality Issues Solution 9.* [sl] (continued)

    1.   Floor drains and sumps that are connected to the drainage system.

    2.   Openings around utility lines (gas and water service lines).

    3.   Hollow concrete block walls.

    4.   Junctions between walls and floors and junctions between floor slab sections.

    5.   In basement floors and cracks in building materials caused by thermal expansion or shrinkage of the materials themselves.

    6.   Exposed soil and rock surrounding the house.

    7.   Unpaved crawl spaces.

10. *Air Quality Issues Solution 10.* [pcy]. In the U.S., most people who are in a room with slow air movement will probably be comfortable at temperatures in the range of 20°C (68°F) to 23.9°C (75°F) during the winter and in the range of 22.8°C (73°F) to 26.1°C (79°F) during the summer.

Humidity is the amount of water vapor within a given space and it is commonly measured as the relative humidity. Relative humidity is defined as the percentage of moisture in the air relative to the amount it could hold if saturated at the same temperature. A relative humidity between 30% and 60% is comfortable to most people.

11. *Air Quality Issues Solution 11.* [sn].

    a.   Toluene percent concentration in the room air is determined as follows:

$$\text{Concentration} = \frac{(\text{Vapor Pressure, mmHg}) (100)}{760} = \frac{(18.4) (100)}{760}$$

$$= 2.42 \text{ gmol\%}$$

Note that this concentration is in units of mole percent. The corresponding mole fraction is 0.0242.

Concentration in ppmv units is determined as follows:

$$\text{Concentration} = \frac{(\text{Vapor Pressure, mmHg}) (1,000,000)}{760}$$

$$= \frac{(18.4) (1,000,000)}{760} = 24,210 \text{ ppmv}$$

11. *Air Quality Issues Solution 11.* [sn] (continued)

From ideal gas law:

$$mg/m^3 \text{ concentration} = \frac{(\text{ppmv concentration}) (MW)}{(0.08205) (273 + °C)}$$

$$= \frac{(24,210) (92.13)}{(0.08205) (293)} = 92,733 \text{ mg/m}^3$$

b. Using the dilution air flow rate equation given in the problem statement, the dilution flow rate for toluene is estimated as follows:

$$W = (0.1) (4 \text{ trays}) (8/12 \text{ ft}) (15/12 \text{ ft}) (2/12 \text{ ft})$$
$$(7.48 \text{ gal/ft}^3) (1/hr)(1 \text{ hr}/60 \text{ min})$$
$$= 6.93 \times 10^{-3} \text{ gal/min}$$

$$Q = \frac{(3222) (0.866) (6.93 \times 10^{-3}) (10) (1,000,000)}{(92.13) (50)} = 41,952 \text{ cfm}$$

c. The volume flow rate of air through the hood is determined as follows:

$$Q_{hood} = (100 \text{ fpm}) (2 \text{ ft}) (4 \text{ ft}) = 800 \text{ cfm}$$

Clearly, there is not enough dilution ventilation to reduce the amount of toluene vapor in the room to an acceptable level so that the student can work safely during the process of immobilization. The best way to ensure a safe workplace is to prevent the toluene vapor from entering the room in the first place by conducting the experiment in a fume hood.

12. *Air Quality Issues Solution 12.* [sl]. A mass balance on VOCs leads to the following equation:

$$\frac{dy}{dt} V_s = Q_L (c_{in} - c) - Q_g (y - y_{in})$$

The equation is rearranged as follows:

$$\frac{dy}{dt} = \frac{Q_L}{V_s} (c_{in} - c) - \frac{Q_g}{V_s} (y - y_{in})$$

12. *Air Quality Issues Solution 12*. [sl] (continued)

Substituting known values from the problem statement results in a value of $Q_L/V_s = 0.008$/min, and $Q_g/V_s = 0.023$/min. The final form of the linear, first-degree differential equation becomes:

$$\frac{dy}{dt} = (0.008) (0.4 - \frac{y}{0.3}) - (0.023) (y - 0.015)$$

This equation is solved to yield a value of $y = 0.54$ mg/L.

13. *Air Quality Issues Solution 13*. [dj]. After consulting any of the references provided in the problem statement, the answers are:

1. It is not a good decision to have a stack that is twice the height of the nearest building or obstacle. It is a better choice to have a stack height that is at least 2.5 times the height of the nearest surrounding building or obstacle so that there is minimal risk of entrainment of the plume into the turbulent wakes of the buildings or obstacles.
2. A factor of two dilution of the stack exit gas will generally have very little effect on the maximum downwind concentration, because the dilution by atmospheric winds is so much larger. Thus, the claim by the incinerator's designers that the concentration will be reduced by half is invalid. Addition of dilution air to the stack may decrease plume concentrations close to the stack, and hence may decrease ground-level concentrations whenever the plume hits the ground near the stack, but the effect of stack dilution air will probably be insignificant compared to atmospheric dilution.

The application for the permit should probably be sent back for stack design revisions before it can be approved.

14. *Air Quality Issues Solution 14*. [jr].

a. The collection efficiency is calculated by:
$$\eta = 1 - \frac{0.04}{4.41} = 0.9909 = 99.09\%$$

b. Substitution into the Deutsch-Anderson equation gives:
$$0.9909 = 1 - \exp\left(-\frac{w\,(60,000)}{5,000}\right)$$

14. *Air Quality Issues Solution 14.* [jr] (continued)

Solving for the drift velocity yields:

$$w = 0.392 \text{ ft/min} = 0.0065 \text{ ft/s}$$

15. *Air Quality Issues Solution 15.* [dks].

a. Linear or non-linear regression techniques can be used to determine first-order mass transfer coefficients for benzene and anthracene using data from Table 16 and the stripping equation provided in the problem statement.
Using data for benzene, the regression shown in Figure 7 was produced.

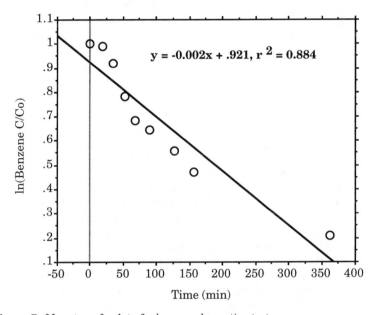

Figure 7. Mass transfer data for benzene desorption test.

From this regression a benzene desorption coefficient was found to be = 0.002/min = 0.12/hr.

b. Using data for anthracene, the regression shown in Figure 8 was produced. From this regression, an anthracene desorption coefficient was found to be = 0.024/min = 1.44/hr.

c. The two results differ by a factor of approximately 10. The main reason for the difference is in the chemical properties of the

15. *Air Quality Issues Solution 15.* [dks] (continued)

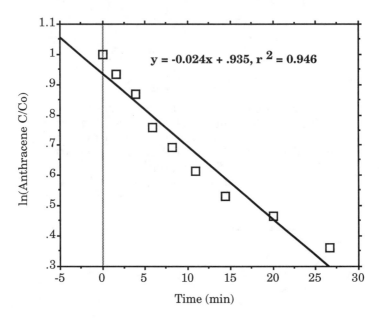

$y = -0.024x + .935, r^2 = 0.946$

Figure 8. Mass transfer data for anthracene desorption test.

two species, primarily the solubility, but second-
arily the diffusivity of these compounds which is
related to their molecular weights.

d. Anthracene is sparingly soluble in water (4.2×10^{-7} M) while benzene has a relatively high
solubility (0.023 M). The mass transfer coef-
ficient of chemical species with a large solubility
is controlled by the turbulence in the gas phase
while that for slightly soluble species is
controlled by the turbulence in the liquid phase.
Thus, because of the high turbulence in the
liquid phase, the rate of anthracene transfer is
high. For benzene, the gas phase turbulence is
low, only controlled by the wind at the low
velocity of 4 mph. So, since the benzene is gas-
phase controlled, the lower turbulence has
reduced its overall mass transfer coefficient.

16. *Air Quality Issues Solution 16.* [kg]. The major control
devices for particulate removal for flue gases are cyclones, fabric
filters (baghouse), electrostatic precipitators, and wet scrubbers.

The major control devices for gaseous contaminant removal
from flue gases are wet scrubbers/venturi scrubbers, packed towers,

16. *Air Quality Issues Solution 16.* [kg] (continued)

adsorption systems, incinerators, condensers and dry scrubbers, and electrostatically augmented fabric filtration systems.

17. *Air Quality Issues Solution 17.* [kg].

    a. The main advantages of electrostatic precipitators are that they have extremely high particulate collection efficiency for coarse and fine particulates at relatively low energy cost, with low pressure drop. They are applicable for high temperature (to 1300°F), large volume flue gases under even high pressure (to 150 psi) or under vacuum conditions.
The main disadvantages of electrostatic precipitators are their high capital cost, their collection efficiency sensitivity to particle characteristics and flowrate fluctuations, their relatively large size, and their potential to form ozone as a by- product.

    b. The main advantages of a fabric filter system are its ability to collect fine particulates with extremely high efficiency, its ability to handle fluctuating flow and fluctuating particulate concentrations, its ability to collect particles as dry materials, its relatively simple operation, and its availability in various configurations.
The main disadvantages are its inability to handle high temperatures, its potential for explosion, its limitations in handling moist acidic or alkaline flue gases, and its requirement for relatively high maintenance.

    c. The main advantages of wet scrubbers are their ability to handle gases with high humidity, high temperature, acidic or alkaline properties, and their ability to achieve high efficiency for fine particulates.
The main disadvantages of wet scrubbers are the potential for creating a water pollution problem, the potential for initiating corrosion problems, their high energy requirements, and their high maintenance costs.

18. *Air Quality Issues Solution 18.* [sn].

Coal requirements = [(500 MW) (1000 kW/MW) (kJ/kW-s) (3600 s/hr) (24 hr/d)]/0.4 = 1.08 x 10^{11} kJ/d

18. *Air Quality Issues Solution 18.* [sn] (continued)

$$= (1.08 \times 10^{11} \text{ kJ/d})/(30{,}000 \text{ kJ/kg}) = 3.6 \times 106 \text{ kg of coal/d}$$

The permitted emission of the various contaminants is:

particulate: $(1.08 \times 10^{11} \text{ kJ/d}) (13 \text{ g/10}^6 \text{ kJ}) (\text{kg/1000 g})$
$$= 1{,}404 \text{ kg/d}$$

NO_x: $(1.08 \times 10^{11} \text{ kJ/d}) (260 \text{ g/10}^6 \text{ kJ}) (\text{kg/1000 g})$
$$= 28{,}080 \text{ kg/d}$$

SO_2: $(1.08 \times 10^{11} \text{ kJ/d}) (260 \text{ g/10}^6 \text{ kJ}) (\text{kg/1000 g})$
$$= 28{,}080 \text{ kg/d}$$

Since there is no permit requirement for CO_2, the rate of emission can be calculated from mass balance considerations:

$CO_2 = (3.6 \times 10^6 \text{ kg coal/d}) (0.772 \text{ kg C/kg coal}) (44 \text{ kg } CO_2/12 \text{ kg C})$
$$= 1.02 \times 10^7 \text{ kg/d}$$

19. *Air Quality Issues Solution 19.* [pcy]. Based on the values given in the problem statement, the plume rise expected at the Grand Gulf power plant can be projected as follows:

$$S = 4 \text{ m, } V_S = 16 \text{ m/s, } U = 5 \text{ m/s}$$

$$T_S = 95 + 273 = 368 \text{ K}$$

$$T_2 = 25 + 273 = 298 \text{ K}$$

$$dT = 368 - 298 = 70 \text{ K}$$

$$dH = 4 \ (16/5)^{1.4} \ (1 + 70/368) = 24.3 \text{ m}$$

20. *Air Quality Issues Solution 20.* [pcy]. Using the Holland plume rise equation and the coefficient values presented in the problem statement, the effective height for the plume is estimated as follows:

$$dH = \left(\frac{(16) \ (4)}{5} \right) \left\{ 1.5 + 2.68 \times 10^{-3} \left(\frac{70}{368} \right) (4) \right\}$$
$$= 12.8 \ (1.5 + 2.14) = 12.8 \ (3.64) = 46.6 \text{ m}$$

21. *Air Quality Issues Solution 21.* [sxl].

   a. The necessary filter surface area is determined from the continuity equation as indicated below:

21. *Air Quality Issues Solution 21.* [sxl] (continued)

$$V = Q/A$$

Rearrangement yields:

$$A = Q/V$$

Based on data for the baghouse given in the problem statement, the required filter area is:

A= (1.7 x 10^6 ft^3/min) (1 min/60 s)/[(0.02 m/s) (3.281 ft/m)
= 432,000 ft^2

b.   The area of one bag is given as:

$$\text{Bag area} = \pi\, D\, L$$

Based on data for the baghouse given in the problem statement, the area for each bag is:

Bag area = π (1 ft) (38 ft) = 120 ft^2

The number of bags required to treat this gas stream based on the area for each bag is:

# of bags required = 432,000 ft^2/120 ft^2 = 3,600 bags

22. *Air Quality Issues Solution 22.* [sxl]. If each bag must be cleaned once per hour with a cleaning time of 3 minutes, the percent of time each bag is out of service is:

Percent of time each bag is out of service = 3 min per hour
= 3 min/60 min = 5%

Therefore 5% more bags must be added to maintain the required surface area of 432,000 ft^2. The number of extra bags required is:

Number of extra bags required = 0.05 x 3,600 = 180 bags

The total number of bags now needed is simply:

Total number of bags now needed = 3,600 + 180 = 3,780 bags

23. *Air Quality Issues Solution 23.* [fwk].   Frequency-of-occurrence data are best handled using log-normal graph paper or via normal score computations with the frequency expressed in cumulative form. That way, a single valued function is being dealt

23. *Air Quality Issues Solution 23.* [fwk] (continued)

with. To convert the data to cumulative form, add to Table 17 a column which shows the percent of time that all listed values are less than the current value as shown below in Table 18.

Table 18
Historical SO_2 Data from Washington, D.C. with
Cumulative Distribution Calculations

1-Hour Average SO_2 Concentration (ppmv)	% Time Concentration Occurred	% Time Concentration or Lower Occurred
0.10 to 0.45	10.0	90.0
0.07 to 0.10	10.0	80.0
0.06 to 0.07	10.0	70.0
0.04 to 0.06	10.0	60.0
0.03 to 0.04	10.0	50.0
0.02 to 0.03	20.0	30.0
0.01 to 0.02	20.0	10.0
0.00 to 0.01	10.0	0.0

A plot of the first versus the third column results in the traditional "s-shaped" cumulative distribution curve. When plotting, a bit of logic is needed to choose the correct concentration from each range listed in the first column. The logic is based on the wording of the third column heading, i.e., the percent of time that the concentration is at or below the stated value. Thus, during the worst 10% of the time, the concentration is at or above 0.10, not 0.45. Similarly, during the worst 20% of the time, the concentration is at or above 0.07, not 0.10. (Can you convince yourself that the average is not an appropriate variable?)

Interpolations, and extrapolations, would be easier if we had a straight line. There are good theoretical arguments, and lots of practical examples, for which the line will be straight if the data were plotted as the log of the concentration versus the number of standard deviations of the frequency, or the standard score for the distribution. Figure 9 shows the log concentration versus normal score for the data contained in Table 18. This plot may also be generated without computation by using log-normal graph paper. The concentrations are plotted on the log scale, and the cumulative frequencies of occurrence on a Gaussian scale. Some spreadsheets support such plots. Note that, while one can shift the decimal point on all values on a log scale, this must never be done on a probability scale.

Once the cumulative values have been calculated, a straight line can be drawn through the data. Interpolation and/or extrapolation of the data can easily be carried out at this point using this linearized relationship as indicated in Figure 9.

23. *Air Quality Issues Solution 23.* [fwk] (continued)

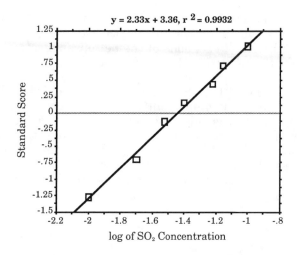

Figure 9. Plot of log concentration versus normal score for the data contained in Table 18.

24. *Air Quality Issues Solution 24.* [dj]. (The solution provided below is an example. Your solution may be completely different. You must, however, do an effective job of explaining to the concerned citizens and the AQCB the difference between physical stack height and effective stack height.)

I am happy to respond to the concerns raised about the height of the proposed stack. This is an excellent question! The proposed stack has a physical height, from base to top, of 110 m. The plume that comes from the stack, under worst-case conditions, will rise another 30 m before it travels horizontally with the wind. You can see good examples of plume rise by observing smoke rising from the chimney connected to a home fireplace. The plume is warmer than the surrounding air, and rises, in much the same manner as a hot air balloon, until it cools off to a temperature that is similar to the temperature of the surrounding air. You may have also seen that plumes from home chimneys tend to bend over closer to the ground in strong winds than they do in light winds. Other common examples of plume rise include rising smoke from campfires or steam rising from a pot of hot soup. Our calculated 30 m rise is a worst-case estimate of how close to the ground the winds at the site may bend the plume.

The sum of the physical stack height, 110 m, and the expected plume rise, 30 m, gives the effective stack height, 140 m, about 460 ft (or one and a half football fields) above the ground surface. The effective stack height is the elevation at which, if we took a big

24. *Air Quality Solution 24.* [dj] (continued).

knife and sliced horizontally, we would split the plume into two halves, much like splitting a hot dog lengthwise. We use the term "effective stack height" in sections of the permit application where we run calculations to predict the down-wind concentrations of the trace gases that may be present in the plume.

So, you can see that the physical stack height and the effective stack height are really two different things, as the effective stack height is the sum of the physical height of the stack plus the expected rise of the plume from the stack. Finally, the positive effect of plume rise is to lift the plume higher off the ground and reduce the strength of the plume at ground level.

Thank you for allowing me to respond to this very viable concern.

Chapter 7

WATER QUALITY ISSUES

Sean X. Liu

I. INTRODUCTION

Since ancient times water has been considered the most suitable medium to clean, disperse, transport, and dispose of waste from human activities. Our use of this resource for culinary, agricultural, industrial, and recreational needs puts an increasingly stringent demand on its management for multiple purposes. With the advent of industrialization and increasing population, the range of requirements for water has expanded as has the demand for higher quality water. Over time, water requirements have emerged for drinking and personal hygiene, fisheries, agriculture (irrigation and livestock supply), navigation for transport of goods, industrial production, cooling in power plants, hydropower generation, and recreational activities such as swimming or fishing. Fortunately, the largest demands for water quantity, such as for agricultural irrigation and industrial cooling, require the least stringent level of water quality. Drinking water supplies and specialized industrial manufacturers exert the most sophisticated demands on water quality but their quantitative needs are relatively moderate.

Water pollution refers to any change in natural waters that may impair further use of the waters, caused by the introduction of organic or inorganic substances, or by a change in temperature of the water. Wastewaters emanate from the following primary sources: municipal wastewater, industrial wastewater, agricultural runoff, and stormwater and urban runoff.

In light of the complexity of factors determining water quality, and the large number of variables involved in describing the status of water bodies in quantitative terms, it is difficult to develop a universally applicable standard which can define the baseline chemical or biological quality for water. Rather, the description of water quality is made either through quantitative measurements, such as physicochemical determinations (particulate matter, biological solids, etc.) and biochemical/biological tests (BOD_5 measurement, toxicity tests, etc.), or through semi-qualitative and qualitative descriptors such as biotic indices, visual aspects, species inventories, odors, etc.

The basic objective of the field of water quality management is to determine the environmental controls that must be instituted to achieve a specific environmental quality objective. A multi-disciplinary approach is usually needed when a receiving water must meet competing agricultural, municipal, recreational, and industrial requirements. In many cases, a cost-benefit ratio must be established between the benefit derived from a specified water quality and the cost of achieving that quality. Many wastewater treatment technologies have been developed for different circumstances and requirements. The most common system practiced in urban areas is the use of wastewater treatment plants based upon a combination of physical, biological and chemical treatment steps. These are generally divided into five consecutive treatment stages:

1. Preliminary treatment. Screening of large material.
2. Primary treatment. Removal of settleable solids, which are separated as sludge for disposal.
3. Secondary (biological) treatment. Here biodegradable organic waste is decomposed by microorganisms.
4. Tertiary (advanced) treatment. Further physical/chemical treatment of secondary effluents, e.g., through chemical coagulation and/or filtration, for the removal of non-biodegradable residual contaminants.
5. Sludge treatment. Dewatering, stabilization and ultimate disposal of sludge generated in previous treatment steps.

In the U.S., municipal treatment plants are required to treat to secondary treatment levels. The use of tertiary treatment is not common but has been increasing based on increasingly stringent stream standards. Increased emphasis has recently been placed upon the removal of secondary pollutants such as nutrients and refractory organics and upon water reuse for industrial and agricultural purposes. This emphasis has generated research, both fundamental and applied, which has improved both the design and operation of wastewater treatment facilities.

In order to attain a water quality management goal, activities must be centered upon the assignment of allowable discharges to a water body so that a designated water use and water quality standards are met using the basic principles of cost-benefit analysis. It is generally not sufficient to simply carry out an engineering analysis of the effect of waste load inputs on water quality. The analysis must also include economic impacts which, in turn, must also recognize the sociopolitical constraints that are operating in the overall problem context.

Water quality standards are usually based on one of two primary criteria: stream standards or effluent standards. Stream standards are based upon receiving-water quality concentrations determined from threshold values of specific pollutants that are required to be maintained in the receiving water to maintain a specific beneficial use of the water into which the waste is being discharged. Effluent standards establish the concentration of pollutants which can be discharged (the maximum concentration of a pollutant, mg/L or the maximum load, lb/d) to a receiving water or upon the degree of treatment required for a given type of wastewater discharge, no matter where in the U.S. the waste originates. These effluent limitations are related to the characteristics of the discharger, not the receiving stream.

It should be noted that the water quality to be attained is not static but is subject to modification with a changing municipal and industrial environment. For example, as the carbonaceous organic load is removed by wastewater treatment, the detrimental effect of nitrification may also become a serious problem in some cases. These considerations may require an upgrading of the degree of treatment provided for waste discharges over time, and suggest that the assessment of water quality conditions and water quality demands must be an on-going process.

The problems in this chapter focus on information gathered from interrelated activities to support a decision regarding what environmental controls or practices must be implemented to achieve a specific level of water quality required for one or more of the competing water demands discussed above. Both point and non-point source pollution are covered along with municipal, agricultural and industrial waste and sludge management issues, water supply and groundwater remediation problems, and regulatory control questions.

II. PROBLEMS

1. *Water Quality Issues Problem 1*. (SDWA, EPA, contaminants). [mh]. In a short, one paragraph essay describe the Safe Drinking Water Act (SDWA) and its goals.

2. *Water Quality Issues Problem 2*. (acid rain, Clean Air Act, Title IV). [hb]. Acid rain results from the dissolution of nitrogen and sulfur oxides into precipitation for form acids that reach the earth's surface. In recent years studies have shown that the pH level of rain has decreased, becoming acidic as the level of these nitrogen and sulfur oxides has increased. Areas in Scotland and the northeastern U.S. have shown the most dramatic effects of this acidic deposition. Acid rain potentially harms forests, lakes, and even drinking water in extreme cases. How did Title IV of the Clean Air Act of 1990 address the acid rain problem in the U.S.?

3. *Water Quality Issues Problem 3*. (eutrophication, lakes, water quality). [car]. What is eutrophication of surface waters? What are the main nutrients of concern in the eutrophication of lakes?

4. *Water Quality Issues Problem 4*. (eutrophication, nutrients, water quality). [car]. What are some methods of reducing the input of nutrients, primarily nitrogen and phosphorus, to the aquatic environment? Both point sources and non-point sources should be considered.

5. *Water Quality Issues Problem 5*. (surface water, water quality, source of inputs). [car]. Answer the following questions related to types of discharges to surface waters.

   a. What are the two major categories of discharges into surface waters? Give an example of each category.
   b. Which of the two major categories would be easier to regulate and why?
   c. Which would be harder to regulate and why?

6. *Water Quality Issues Problem 6*. (primary treatment, secondary treatment, advanced wastewater treatment, municipal wastewater treatment). [mh]. Discuss in general terms what comprises municipal wastewater treatment.

7. *Water Quality Issues Problem 7*. (water and wastewater treatment, water quality assessment, BOD, COD, TOC, TOD). [sl]. Define the following terms used as water quality descriptors and briefly explain their differences related to the methods used for their determination.

7. *Water Quality Issues Problem 7.* [sl] (continued)

    a. COD
    b. TOC
    c. TOD
    d. BOD

8. *Water Quality Issues Problem 8.* (wastewater treatment, BOD_5 discharges). [sl]. A typical city of 45,000 people has a wastewater treatment discharge of 5.5 million gal/d (MGD) and a BOD_5 (5-day biochemical oxygen demand) in the raw wastewater of 175 mg/L. What is the total discharge of BOD in lb/d? What is the BOD_5 discharge in lb/person/d?

9. *Water Quality Issues Problem 9.* (sludge management, water treatment). [sl]. Sludge generated in a water or wastewater treatment plant usually contains substantial amounts of water and therefore needs to be processed to reduce the water content (the process is called sludge dewatering) for ultimate disposal or landfilling. A municipal water treatment plant in Kansas produces 1.05 tons of sludge every day. The wet sludge (before dewatering) has a density of 1.05 g/cm^3, which increases to 1.65 g/cm^3 after being treated. How much additional space would be needed in a landfill site annually if the wet sludge was dumped directly into the landfill site without dewatering? Note: this practice is no longer permitted by law because of leachate and gas production concerns in sanitary landfills.

10. *Water Quality Issues Problem 10.* (flocculation, agglomeration optimum-dose, water treatment). [sl]. Flocculation is a physical process used to encourage small particles to aggregate into larger particles or floc. It is an essential component of most water treatment plants in which flocculation, sedimentation, and filtration processes are integrated to effectively remove suspended particles from water. Chemicals (such as alum, polyelectrolytes, etc.) are usually added to achieve agglomeration among small particles in water.

In order to determine the optimum chemical dose for flocculation, jar tests are used to estimate the amount of chemicals needed to ensure proper flocculation. In a jar test conducted for a given water treatment plant, 4-L samples were poured into a series of jars. After the test, the jar which has been dosed with 20 mL of alum solution containing 5 mg Al(III)/mL showed optimal results. If the amount of water to be treated is 90 MGD, calculate the pounds per day of alum [$Al_2(SO_4)_3 \bullet 14H_2O$, MW = 594.4 g/gmol] that should be added to the raw water.

11. *Water Quality Issues Problem 11.* (Cyanide waste, stoichiometry). [sn]. Cyanide-bearing waste is to be treated by a batch process using alkaline chlorination. In this process, cyanide is reacted with chlorine under alkaline conditions to produce carbon dioxide and nitrogen as end-products. The cyanide holding tank contains 28 m^3 with a cyanide concentration of 18 mg/L. Assuming that the reaction proceeds according to its stoichiometry, answer the following questions.

    a. How many pounds of chlorine are needed?
    b. How long will the hypochlorinator have to operate if the hypochlorinator can deliver 900 L/d of chlorine?
    c. How long should the caustic soda feed pump operate if the pump delivers 900 L/d of 10 wt% caustic soda solution?

12. *Water Quality Issues Problem 12.* (self-purification). [sn]. Natural water bodies possess a capacity to stabilize organic matter without seriously affecting their general quality and aesthetics. The process is called self-purification which involves various microorganisms (e.g., bacteria, algae, etc.). Draw a diagram incorporating the processes that could occur in a lake or pond when wastewater is discharged intermittently into it. Describe the carbon and nitrogen cycles related to these processes and locate where they should occur in the water body.

13. *Water Quality Issues Problem 13.* (microbial regrowth, water distribution system). [sn]. Microbial regrowth can be defined as an increase in viable microorganism concentrations in drinking water down stream of the point of disinfection after treatment. These microorganisms may be coliform bacteria, bacteria enumerated by the heterotrophic plate count (HPC bacteria), other bacteria, fungi, or yeasts. Regrowth of bacteria in drinking water can lead to numerous associated problems including multiplication of pathogenic bacteria such as *Legionella pneumophila*, deterioration of taste, odor, and color of treated water, and intensified degradation of the water mains, particularly cast iron, by creating anaerobic conditions and reducing pH in a limited area. In order to obtain stable drinking water (i.e., to control regrowth), one needs to understand the sources of biological instability. Explore the possible factors that affect regrowth in a water distribution system.

14. *Water Quality Issues Problem 14.* (groundwater treatment, total dissolved solids, evaporator). [sl]. A preliminary evaluation of a contaminated groundwater site, located near Nowhere, Utah, shows that the main contaminant at the site is total dissolved solids

14. *Water Quality Issues Problem 14.* [sl] (continued)

(TDS). The TDS concentrations, consisting mainly of sulfates, nitrates, and sodium, were found to be in the 14,000 to 20,000 mg/L range. TDS tends to diffuse into the bulk phase of groundwater and contaminates the drinking water source as the groundwater flows through a high salt content formation. Most water treatment processes are not capable of effectively removing this TDS which causes odor, color, taste problems, and even cancer.

A team of experts was assembled to tackle the problem. Various technologies were considered for remediation of the groundwater at the site. In the end evaporators were chosen to accomplish the task because of the high concentration of TDS in the contaminated groundwater. Evaporators use a heat source to concentrate a solution or to recover dissolved solids by evaporating off the water and then returning the water back to the aquifer after it is condensed as pure water.

A pilot study was conducted to verify that this technology would be effective at this site. The evaporator used for the pilot test had a capacity of 50,000 gallons per day (gpd) and operated a total of 15 days and 3 hours during the pilot study. The target TDS concentration was set at 200 mg/L. Groundwater with a final TDS concentration below this target level was considered remediated. If it is assumed that the average value of the concentration range is representative of the initial TDS concentration (17,000 mg/L), what amount of solids was removed from the site during the pilot test treatment of this groundwater?

15. *Water Quality Issues Problem 15.* (toxic, chromium). [pcy]. List the possible sources of highly toxic hexavalent chromium (Cr^{6+}) and methods to remove it from a wastewater stream.

16. *Water Quality Issues Problem 16.* (industrial wastewater treatment, land treatment, BOD reduction). [sl]. Land treatment of industrial wastewater is a process in which wastewater is applied directly to the land. This type of treatment is most common for food processing wastewater including meat, poultry, dairy, brewery, and winery wastes. The principal rationale of this practice is that the soil is a highly efficient biological treatment reactor, and food processing wastewater is highly degradable. This treatment practice is usually carried out by distributing the wastewater through spray nozzles onto the land or letting the water run through irrigation channels.

Suppose that the rate of the wastewater flowing to a land application site is 178 gal/acre/min and the irrigated land area is 5.63 acres. The entire irrigation process lasts 7.5 hr/d. If the wastewater has a BOD_5 concentration of 50 mg/L, what is the mass of BOD_5 remaining in the soil after the land treatment process

16. *Water Quality Issues Problem 16.* [sl] (continued)

is complete (assume a BOD_5 removal efficiency by the land treatment process is 95%)?

17. *Water Quality Issues Problem 17.* (chlorine, disinfection, economics, ultraviolet light). [dj]. A large wastewater treatment facility currently uses chlorine for disinfection. The average chlorine dosage is 6.0 mg/L at an average flow rate of 70 MGD. Chlorine (Cl_2) costs $1.00/lb. Sulfur dioxide (SO_2) is used to dechlorinate the effluent before it is discharged (requirement placed on the facility to protect the fish inhabiting the receiving water) and is consumed at an average dose of 2.0 mg/L. Its cost is $1.20/lb.

Concerned about operating costs and the risks of Cl_2 usage, the utility has completed a pilot-study showing that ultraviolet (UV) light disinfection could be used to replace Cl_2 and SO_2. The UV system will have a capital cost of $12,000,000, will cost $500,000 a year to operate, and will have a 20-year service life. The utility will draw from bank savings accounts currently earning 8.0% simple interest per year to pay for the UV system's capital cost. Does the savings in operating costs justify the capital cost? Assume straight-line depreciation of the UV system, with $0 salvage value at the end of its service life.

Needed information:

Conversion factor:
A dose of 1 mg/L = 8.34 lb/10^6 gal (MG)

Straight-line depreciation equation:

$$\text{Depreciation rate in \$/yr} = \frac{\text{(initial value - salvage value)}}{\text{(service life)}}$$

Rate of return on investment (ROI) equation in %/yr:

$$ROI = \frac{\text{(Scenario A operating costs) - (Scenario B operating costs)}}{\text{(investment)}} (100)$$

where Scenario A = the estimated cost for continuing the status quo, and Scenario B = the estimated cost for the proposed alternative.

18. *Water Quality Issues Problem 18.* (water and wastewater treatment, adsorption, pollutant removal). [sl]. Adsorption processes are widely used in water and wastewater treatment to remove organic molecules that cause taste, odor, color and toxicity. The relationship between adsorbents (solid materials that adsorb organic matter, e.g., activated carbon) and adsorbates (substances that are bound to the adsorbates, e.g., benzene) may be simply expressed as:

$$q = Kc^n$$

where q = the amount of organic matter adsorbed per amount of adsorbent; c = the concentration of organic matter in water; n = an experimentally determined constant; and K = an equilibrium distribution constant. A college student who worked for a company as an intern was asked to evaluate the K and n values for a certain type of activated carbon based upon the data given in the table below that were obtained from a laboratory sorption experiment. What are the values of n and K that the intern should have generated?

Table 19
Sorption Data Collected in Laboratory Experiments

Concentration (mg/L)	q (mg/g)
50	0.118
100	0.316
200	0.894
300	1.64
400	2.53
650	5.24

19. *Water Quality Issues Problem 19.* (water treatment, water quality). [car]. What are some of the water quality issues that affect the operation of water treatment plants for treating domestic water supplies?

20. *Water Quality Issues Problem 20.* (water treatment, water quality). [car]. What are the three major process components used to treat water sources to be used for potable water supplies, and why are they utilized?

21. *Water Quality Issues Problem 21.* (waste load allocation, water quality). [car]. One important aspect of water quality management is the assignment of allowable point source wastewater discharges to a receiving water body such that the water quality objectives for that water body can be maintained. This management process is called Waste Load Allocation (WLA).

21. *Water Quality Issues Problem 21.* [car] (continued)

What are some of the important steps that should be included in the WLA process?

22. *Water Quality Issues Problem 22.* (NPDES, discharge limits, standards, point source, water quality). [car]. Answer the following questions regarding the NPDES program.

    a. What does the acronym NPDES stand for?
    b. What are the three major parameters regulated under the NPDES program for municipal wastewater discharges and what are the maximum concentrations allowed for each of these parameters in NPDES permits?

23. *Water Quality Issues Problem 23.* (wastewater, NPDES, discharge limits, water quality). [car]. The following 5-day biochemical oxygen demand (BOD_5) and total suspended solids (TSS) data were collected at a local municipal wastewater treatment plant over a 7-day period. The NPDES permit limitations for BOD_5 and TSS effluent concentrations from this wastewater treatment plant are 45 mg/L on a 7-day average. Based on this information, is the treatment plant within its NPDES permit limits?

Table 20
Daily BOD_5 and TSS Effluent Concentration Data Collected Over a 7-Day Period at a Municipal Wastewater Treatment Plant

Day	BOD (mg/L)	TSS (mg/L)
1	45	20
2	79	100
3	64	50
4	50	42
5	30	33
6	25	25
7	21	15

24. *Water Quality Issues Problem 24.* (decision objectives, water resource management, water quality). [car]. When determining water quality objectives for a water resource, what are the major information items that should be evaluated?

25. *Water Quality Issues Problem 25.* (groundwater, retardation factor, pore water velocity, contaminant velocity). [dks]. The velocity of a contaminant in groundwater is slowed by the presence of organic matter in the soil into which the contaminant

25. *Water Quality Issues Problem 25.* [dks] (continued)

will partition. Under equilibrium partitioning conditions, contaminant velocity is related to the pore water velocity by

$$v_c = v/R$$

where v_c = the contaminant velocity, length/time; v = the pore water velocity, length/time; and R = the retardation factor, unitless.

For naphthalene in a particular aquifer, R has been found experimentally to equal 70. If the pore water velocity from a source to a well at a distance of 1,000 m is 5×10^{-3} cm/s, what is the travel time of naphthalene to the well?

26. *Water Quality Issues Problem 26.* (groundwater, benzene, underground storage tank, retardation factor, biodegradation, product recovery). [dks]. A total of 400 L of pure benzene leaks from an underground storage tank before the leak is discovered. The water table lies a few feet below the tank. Discuss the following items related to this release:

    a. What is the possibility of recovering some of the pure product benzene, and how might this product recovery be accomplished.

    b. What is the maximum benzene concentration expected in the groundwater?

    c. What is the dissolved benzene retardation factor assuming that the soil organic carbon fraction (f_{oc}) = 0.5%?

    d. What is the rate of biodegradation expected for this benzene?

    NOTE: $R = 1 + f_{oc} (K_{oc}) (\rho_b)/n$, where f_{oc} = the fraction of organic carbon in the soil; K_{oc} = the organic carbon normalized soil/water partition coefficient; ρ_b = the bulk density of the aquifer solids; and n = the aquifer solid total porosity. For benzene, K_{oc} is reasonably approximated by its octanol/water partition coefficient, $K_{ow} \approx 100$ (mL water/g octanol). Typical values of n and ρ_b for aquifer solids are 0.3 and 2 g/mL, respectively.

27. *Water Quality Issues Problem 27.* (groundwater, ammonia, animal waste management, non-point source). [dks]. Echo Creek runs through a cattle ranch in northern Utah and is used as drinking water for the cattle. Due to the fact that cattle indiscriminately dispose of their waste products on the soil, Echo

27. *Water Quality Issues Problem 27.* [dks] (continued)

Creek's riparian zone, which extends from the creek's bank 10 m on either side, for 1 km of the channel, has been approximately, uniformly contaminated with manure.

Runoff (with an average depth of 0.5 cm) from precipitation flows perpendicularly through the riparian zone toward the stream at a velocity of 12 m/d (precipitation falling within the riparian zone may be ignored for your calculations). This runoff leaches organic contaminants and ammonia (NH_3) from the manure at a first order rate that is proportional to the difference between the contaminant concentration in the water and their ultimate water solubilities. The ultimate water solubilities of the organic matter and ammonia are 2,000 g/m^3 and 10,000 g total ammonia (all species)/m^3, respectively. The leaching rate constant for both contaminants is 0.30 d^{-1}.

It may be assumed that there is no mixing of contaminants in the direction of, or perpendicular to the direction of flow, i.e., the stream may be assumed to be a plug flow system. The flow in the stream above the ranch is 0.3 m^3/s. Downstream of the ranch, water quality samples obtained during a storm indicate an organic contaminant concentration of 6 ± 2 mg/L (2 is the standard error of the mean of 10 measurements).

Above the ranch, the stream pH is 7.0 and the total alkalinity is 75 mg/L as $CaCO_3$. The runoff can be assumed to have the characteristics of rain water before entering the riparian zone, that is, it is saturated with CO_2, O_2, and N_2, but all other concentrations are zero.

Based on the information given above, answer the following questions. Explain your answers, stating all assumptions and show any calculations that may be required.

    a. Do you think that the ranch is the only important contributor of the contaminant to Echo Creek? Support your answer with calculations.

    b. What is the concentration of unionized ammonia (NH_3(aq)) at the downstream end of the ranch? The pKa for $NH_4^+ = NH_3 + H^+$ is 9.3 at the stream temperature of 20°C.

III. SOLUTIONS

1. *Water Quality Issues Solution 1*. [mh]. The Safe Drinking Water Act was created in 1974 to protect the quality and safety of public drinking water supplies. Under the control of the U.S. EPA, the Act created national standards for acceptable levels of contaminants that may be tolerated in drinking water. By 1986, the U.S. EPA identified 83 known contaminants in the nation's water supplies. The Act also regulates underground injection wells and protects sole source aquifers.

2. *Water Quality Issues Solution 2*. [mh]. Title IV mandates controls to reduce the emissions of sulfur and nitrogen oxides. The Act requires 110 fossil fuel fired power plants to reduce their emissions rate to 1985 to 1987 levels (2.5 lb of SO_2/MMBtu times the average of their 1985 to 1987 fuel usage). The second part of the Act requires approximately 2,000 utilities to reduce their emissions to a level equivalent to the product of an emission rate of 1.2 lb SO_2/MMBtu times the average of their 1985 to 1987 fuel use.

In order to assist areas that will have difficulty reaching these mandated emission levels, those areas which are not in jeopardy of violating Clean Air Act standards may sell shares of their air pollutant loading to other emitters. Finally, continuous emissions monitoring devices must be installed on all affected sources, and a compliance plan must be approved as well.

3. *Water Quality Issues Solution 3*. [car]. Eutrophication of lakes is the over production of biomass, primarily algae and bacteria, due to excess nutrient loading to them from untreated or partially treated wastewater discharges or non-point runoff. Limited nutrient concentrations typically limit excessive algae growth. When nutrients are added to a system in excess, algal blooms, excess biomass, excessive bacterial growth, oxygen depletion and water quality deterioration follow. This process of algal blooms and subsequent environmental deterioration is termed eutrophication.

The two major nutrients that, when their concentrations are increased in the aquatic environment, can result in eutrophication are nitrogen and phosphorus.

The general forms these nutrients can be found in within the aquatic environment are:

- inorganic nitrogen, such as ammonia, nitrite, and nitrate;
- organic nitrogen, which is bound up as protein material or in organic fertilizer complexes;
- inorganic phosphorus, such as calcium or potassium phosphate, both of which are dissolved from soil; and

3. *Water Quality Issues Solution 3.* [car] (continued)

- organic phosphorus, which is bound up in fertilizer complexes, or in lipid material.

4. *Water Quality Issues Solution 4.* [car]. Some methods of reducing the input of nitrogen and phosphorus into the aquatic environment include:

- Treating point sources, such as municipal wastewater, storm water runoff and industrial wastewater, utilizing biological, chemical and physical treatment processes.
- Utilizing land conservation practices, such as buffer strips and erosion control measures, to limit the runoff potential of these nutrients from agricultural land.
- Reducing the amount of fertilizer application by monitoring concentrations in the soil available for plants to use or by utilizing sustainable agricultural methods.
- Limiting the input of nutrient rich wastewater discharges into receiving streams by recycling the wastewater, such as using the wastewater for irrigation of parks, golf courses or cemeteries.
- Reducing the input of nutrients into wastewater by modification of the product, such as use of non-phosphate detergents.

5. *Water Quality Issues Solution 5.* [car]

a. The two major categories of discharges into surface waters are:
  - point sources which are essentially any discharge that can be specifically located by finding the discharge pipe. A few examples include:
    - industrial discharges
    - municipal wastewater discharges
    - municipal storm water discharges
  - non-point sources. A few examples include:
    - rainfall runoff from large areas of land, as in agricultural areas, large fields, etc.
    - atmospheric deposition

b. The easiest category of discharge source to regulate is the point source. The point source essentially has a physical structure that extends from the source of the discharge to the receiving

5. *Water Quality Issues Solution 5.* [car] (continued)

body or surface water. Thus, there is one point or location that can be sampled to check for compliance with set discharge standards.

c. The hardest category of discharge source to regulate is the non-point source. This source of discharge covers a wide and varying area and a sample cannot be taken that is truly representative of a specific location. Thus, there are numerous variables that could impact the discharge sample that limits the enforcement of a set of discharge standards for this type of source.

6. *Water Quality Issues Solution 6.* [mh]. Wastewater treatment plants are designed to provide either secondary or advanced (tertiary) treatment, depending on the degree of purification required by the water body receiving the treated wastewater. Most secondary and tertiary treatment plants incorporate primary treatment as part of their overall process. Primary treatment utilizes physical processes, such as screening, sedimentation, and filtration, to remove a portion of the pollutants that will settle, float, or that are too large to pass through simple screening devices. These may include paper, rags, old shoes, or even sticks.

The main purpose of secondary treatment is to provide BOD_5 removal beyond what is achievable by simple sedimentation. There are three commonly used approaches, all of which take advantage of the ability of microorganisms to convert organic wastes into stabilized, low energy compounds. Two of these approaches, the trickling filter and the activated sludge processes, sequentially follow normal primary treatment. The third approach, oxidation ponds, can provide equivalent results without primary treatment.

Advanced wastewater treatment is designed to remove those constituents that may not be adequately removed by secondary treatment plants such as nitrogen, phosphorus, and heavy metals.

Sludge is produced as a waste product during the wastewater treatment process, and the management and proper disposal of this solid material can be equivalent in cost and effort to that associated with the treatment of wastewater liquid. Sludge management and disposal includes stabilization (either chemically or biologically), dewatering (by vacuum filters, filter presses, centrifuges or sand drying beds), and ultimate disposal (through land application, landfilling, incineration and ash disposal, or ocean disposal).

7. *Water Quality Issues Solution 7.* [sl].

a. COD = chemical oxygen demand
b. TOC = total organic carbon

7. *Water Quality Issues Solution 7.* [sl] (continued)

    c.   TOD = total oxygen demand
    d.   BOD = biochemical oxygen demand

The COD test is a chemical test; the TOC test is an instrumental test; the TOD represents the theoretical amount of oxygen required to oxidize the waste material completely to oxidized end-products, i.e., CO_2, H_2O, SO_4^{2-}, etc.; and the BOD test is a biochemical test involving the use of microorganisms.

8. *Water Quality Issues Solution 8.* [sl]. The total BOD_5 produced per day is:

$$\text{Total } BOD_5/d = (5.5 \text{ MGD}) (175 \text{ mg/L}) (8.34 \text{ lb/MG/mg/L})$$
$$= 8{,}027 \text{ lb } BOD_5/day$$

The BOD_5 discharge in lb/person/d is:

$$\text{lb } BOD_5/\text{person/d} = (8{,}027 \text{ lb } BOD_5/day)/(45{,}000 \text{ people})$$
$$= 0.18 \text{ lb } BOD_5/\text{person/d}$$

9. *Water Quality Issues Solution 9.* [sl]. The change of volume of sludge before and after dewatering can be expressed in terms of the ratio of densities of the materials as follows:

$$V_{wet}/V_{treated} = \text{treated density/wet density} = 1.65/1.05 = 1.57$$

The decrease of sludge volume is, therefore, 57.1%.

The increase of volume would be:

$$(0.571) (365 \text{ d/yr}) (1.05 \text{ T/d}) (2{,}000 \text{ lb/T}) (454 \text{ g/lb})/(1.65 \text{ g/cm}^3)$$
$$= 120{,}425{,}976 \text{ cm}^3/yr = 120{,}426 \text{ L/year}$$
$$= 120{,}426 \text{ L/year} (1 \text{ gal}/3.785 \text{ L}) (1 \text{ ft}^3/7.48 \text{ gal}) = 4{,}254 \text{ ft}^3/yr$$

The problem may alternatively be solved as follows:

The volume of wet sludge before dewatering:

$$(1.05 \text{ T/d}) (2{,}000 \text{ lb/T}) (454 \text{ g/lb})/(1.05 \text{ g/cm}^3)$$
$$= 908{,}000 \text{ cm}^3/d = 908 \text{ L/d}$$
$$= 908 \text{ L/d} (1 \text{ gal}/3.785 \text{ L}) (1 \text{ ft}^3/7.48 \text{ gal}) = 32.1 \text{ ft}^3/d$$

The volume of treated sludge after dewatering:

$$(1.05 \text{ T/d}) (2{,}000 \text{ lb/T}) (454 \text{ g/lb})/(1.65 \text{ g/cm}^3)$$
$$= 577{,}818 \text{ cm}^3/d = 578 \text{ L/d}$$
$$= 578 \text{ L/d} (1 \text{ gal}/3.785 \text{ L}) (1 \text{ ft}^3/7.48 \text{ gal}) = 20.4 \text{ ft}^3/d$$

9. *Water Quality Issues Solution 9*. [sl] (continued)

The additional volume required in the landfill if the sludge is not dewatered is:

$$(32.1 - 20.4) = 11.7 \text{ ft}^3/\text{d} (365 \text{ d/yr}) = 4{,}271 \text{ ft}^3/\text{yr}$$

10. *Water Quality Issues Solution 10*. [sl]. The dose administrated in the laboratory, neglecting dilution, is:

Alum dose = [(5 mg Al(III)/mL alum) (20 mL alum) (594.4 g/gmol alum)]/[(27 g/gmol Al(III)) (2 gmol Al(III)/gmol alum) (4 L)]
= 275 mg alum/L

Note that 1 gmol alum contains 2 gmol Al(III) ions.

The dosage applied at the plant is:

Alum dosage = (90 MGD) (275 mg alum/L) (8.34 lb/MG/mg/L)
= 206,415 lb/d.

11. *Water Quality Issues Solution 11*. [sn]

a. The balanced equation can be written as follows:

$$2 \text{ CN}^- + 8 \text{ OH}^- \rightarrow 2 \text{ CO}_2 + \text{N}_2 + 4 \text{ H}_2\text{O} + 10 \text{ e}^-$$
$$5 \text{ Cl}_2 + 10 \text{ e}^- \rightarrow 10 \text{ Cl}^-$$

$$2 \text{ CN}^- + 8 \text{ OH}^- + 5 \text{ Cl}_2 \rightarrow 2 \text{ CO}_2 + \text{N}_2 + 10 \text{ Cl}^- + 4 \text{ H}_2\text{O}$$

Therefore, 8 gmol of caustic and 5 gmol of chlorine are required to oxidize 2 gmol of cyanide to carbon dioxide and nitrogen gas.

Calculate the amount of cyanide to be treated.

Cyanide, kg = (CN$^-$, m^3) (CN$^-$, mg/L) (1,000 L/m^3) (1 g/1,000 mg)
= (28 m^3) (18 mg/L) = 504 g cyanide

Calculate the amount of chlorine needed to oxidize this amount of cyanide:

Chlorine, g = (CN$^-$, g) (Cl$_2$ Dose, g/g CN$^-$) (MW of Cl$_2$/MW of CN$^-$)
= [(504 g CN$^-$) (5 gmol Cl$_2$/(2 gmol CN$^-$)] [(71 g/gmol)/(26 g/gmol)]
= 3,441 g Cl$_2$

11. *Water Quality Issues Solution 11.* [sn] (continued)

    b. Calculate the time of operation of the hypochlorinator in hours:

Time, hr = (Cl_2 required, g) (100%) (24 hr/d)/[(hypochlorite flow, m^3/d) (Cl_2 in hypochlorite %)]
= [(3,441 g) (100%) (24 hr/d)]/[(900 L/d) (2%) (1,000 g/L)]
= 4.59 hr = 4 hr 35 min

    c. Calculate the time of operation of the caustic soda pump:

Caustic soda required, g = (CN^-, g) (NaOH Dose, gmol/gmol CN^-) (MW NaOH/MW CN^-)
= [(504 g CN^-) (8 gmol NaOH)/(2 gmol CN^-)] [(40 g/gmol)/(26 g/gmol)]
= 3,102 g NaOH

Time, hr = (NaOH required, g) (100%) (24 hr/d)/[(NaOH flow, L/d) (NaOH concentration,%) (1,000 g/L)]
= [(3,102 g) (100%) (24 hr/d)]/[(900 L/d) (10%) (1,000 g/L)]
= 0.83 hr = 50 min

12. *Water Quality Issues Solution 12.* [sn]. The processes and biological reactions involved in self-purification in a surface water body such as a lake or pond can be incorporated into a diagram as shown in Figure 10.

In general, the water body can be divided into three distinct regions: aerobic (top), facultative (intermediate), and anaerobic (bottom) zones. Aerobic conditions exist near the surface and throughout the water column depth where aeration and photosynthesis take place. Oxygen released by algae is used by the aerobic and facultative bacteria in the stabilization of soluble and colloidal organic matter. The carbon dioxide released in this degradation process is in turn used by the algae for their synthesis of cell material. These symbiotic reactions are described qualitatively in the following equations:

Bacterial Respiration:

    Organic matter + O_2 → New cells + CO_2+ H_2O + Energy

Algal Photosynthesis:

CO_2 + Nutrients + Sunlight → New cells + O_2 + H_2O + Energy

In the presence of sulfide, carbon dioxide that is released by bacteria may also be used by photosynthetic sulfur bacteria as follows:

12. *Water Quality Issues Solution 12.* [sn] (continued)

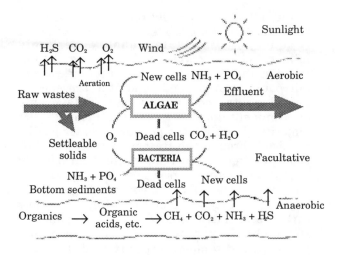

Figure 10. Algal and bacterial activity in a lake or facultative pond.

$$\text{Sunlight} + CO_2 + 2\,H_2S \rightarrow (CH_2O) + 2\,S^- + H_2O + \text{Energy}$$
$$\text{Sunlight} + 3\,CO_2 + 2\,S^- + 5\,H_2O \rightarrow 3\,(CH_2O) + 2\,H_2SO_4 + \text{Energy}$$

At the bottom of the water body, anaerobic conditions may prevail. In this region, solids in the sludge layer are broken down anaerobically to dissolved organics (volatile acids) and gaseous end-products such as CH_4, CO_2, and H_2S.

$$\text{Organic material} + \text{Acid formers} \rightarrow \text{Volatile acids} + \text{Energy}$$
$$\text{Volatile acids} + \text{Methane formers} \rightarrow CH_4 + CO_2 + \text{Energy}$$

Organic nitrogen in the waste is converted to ammonia (ammonification) which is readily used by algae and bacteria for new cell synthesis. When cells die and break down, the nutrient content within the cell is released and recycled to the system. A fraction of the organic nitrogen and phosphorous may be bound in the sediment associated with biomass.

Because of aerobic/anaerobic conditions within the water body, nitrification and denitrification are likely since conditions are favorable for such a series of reactions. At a relatively high pH and temperature, ammonia in the water may be lost through volatilization. Nitrate can also be taken up by algae.

13. *Water Quality Issues Solution 13.* [sn]. The following items are some factors that can affect microbial regrowth in a culinary water distribution system:

13. *Water Quality Issues Solution 13*. [sn] (continued)

a. Water quality (e.g., concentration of organic
carbon and nutrients, temperature, disinfectant
residual, etc.).   Water quality certainly
influences the potential for bacterial regrowth in
the distribution system.   The proliferation of
microorganisms depends upon the presence of
an adequate food supply and a favorable
environment for their growth.   Though the
majority of organic compounds in water can be
removed through the treatment processes, low
concentrations may remain that sufficiently
stimulate bacterial growth.   Some groups of
bacteria (generally known as oligotrophic
bacteria) are capable of surviving under low
carbon and nutrient conditions. Examples of
water quality parameters affecting microbial
regrowth are temperature, pH, and dissolved
oxygen. In general disinfectant residual is used
to suppress growth during storage and
transportation of potable water.

b. Pipe materials and conditions.   The physical
condition of distribution system pipes influences
their tendency to foster biological regrowth.
Distribution pipes that have tubercules or other
surface irregularities commonly harbor microbial encrustations. Certain pipe materials and
conditions can lead to a heavy accumulation of
bacteria on their walls, a so-called biofilm. By
attaching to the surface, microorganisms can be
protected from washout and can exploit larger
nutrient resources either accumulated at the
surface or in the passing water.   Moreover,
attached bacteria appear to be less affected by
disinfectants than those suspended in the
disinfected liquid.

c. Flow conditions in the distribution system (e.g.,
detention time, flow velocity, wall shear stress,
flow reversals, etc.). Long detention times and
low flow velocities tend to encourage bacterial
growth in the distribution system.

d. Water treatment processes. The efficiency of
organic compound and nutrient removal
depends upon the type of treatment processes
utilized at a given plant. Generally, residual
disinfectant in the water is required to prevent
regrowth in the distribution system. Ozonation
is known to increase bacterial nutrients by con-

13. *Water Quality Issues Solution 13*. [sn] (continued)

verting some of the non-biodegradable organic matter into oxidized, degradable compounds, making it more available for uptake by microorganisms colonizing the distribution system.

e. Presence, type, concentration, and physiological state of the bacteria. It is evident that bacteria or other organisms must be present in a distribution system for biological regrowth to occur. The possible sources of bacteria in the water are: 1) recovery of injured or dormant bacteria from disinfection; 2) cross-connections which allow back siphonage of contaminated water; 3) improperly protected distribution system storage; and 4) line breaks and subsequent repair operations that fail to re-disinfect the repaired lines effectively before placing them back into service.

14. *Water Quality Issues Solution 14*. [sl]. The total amount of groundwater treated was:

$$(50,000 \text{ gal/d}) (15.125 \text{ d}) = 756,250 \text{ gal}$$

The total amount of solids removed from this treated groundwater is:

$$(17,000 \text{ mg/L} - 200 \text{ mg/L}) (756,250 \text{ gal}) (3.785 \text{ L/gal})$$
$$= 48,088,425,000 \text{ mg} = 48,088,425 \text{ g}$$
$$= (48,088,425 \text{ g})/(454 \text{ g/lb}) = 105,922 \text{ lb}$$

15. *Water Quality Issues Solution 15*. [pcy]. Hexavalent chromium-bearing wastewater is produced in chromium electro-plating, chromium conversion coating, etching with chromic acid, and in metal finishing operations carried out on chromium as the base material.

Chromium wastes are commonly treated in a two-stage batch process. The primary stage is used to reduce the highly toxic hexavalent chromium to the less toxic trivalent chromium. There are several ways to reduce the hexavalent chrome to trivalent chrome including with sulfur dioxide, bisulfite or ferrous sulfate. The trivalent chrome is then removed by hydroxide precipitation. Most processes use caustic soda ($NaOH$) to precipitate chromium hydroxide. Hydrated lime ($Ca(OH)_2$) may also be used. The chemistry of the reactions is described as follows:

15. *Water Quality Issues Solution 15.* [pcy] (continued)

    a.   Using SO_2, for Cr^{6+} to Cr^{3+}

$$SO_2 + H_2O \rightarrow H_2SO_3$$

$$3\ H_2SO_3 + 2\ H_2CrO_4 \rightarrow Cr_2(SO_4)_3 + 5\ H_2O$$

    b.   Using NaOH to precipitate

$$6\ NaOH + Cr_2(SO_4)_3 \rightarrow 2\ Cr(OH)_3 + 3\ Na_2SO_4$$

16. *Water Quality Issues Solution 16.* [sl]. The amount of the wastewater flowing to the land is:

$$(178\ gal/acre/min)\ (60\ min/hr)\ (7.5\ hr/d)\ (5.63\ acre)$$
$$= 450{,}963\ gal/d$$

The amount of BOD_5 remaining is

$$(450{,}963\ gal/d)\ (1.00 - 0.95)\ (50\ mg/L)\ (3.785\ L/gal) = 4{,}267{,}688\ mg/d$$
$$= 4{,}267{,}688\ mg/d\ (1\ g/1{,}000\ mg)\ (454\ g/lb) = 9.4\ lb/d$$

17. *Water Quality Issues Solution 17.* [dj]. Calculate the daily Cl_2 and SO_2 dosage rates as follows:

$$Cl_2\ dosage\ rate = (flow\ rate)\ (Cl_2\ dose)\ (8.34\ lb/MG/mg/L)$$
$$= (70\ MGD)\ (6.0\ mg/L)\ (8.34\ lb/MG/mg/L) = 3{,}500\ lb/d$$

$$SO_2\ dosage\ rate = (flow\ rate)\ (SO_2\ dose)\ (8.34\ lb/MG/mg/L)$$
$$= (70\ MGD)\ (2.0\ mg/L)\ (8.34\ lb/MG/mg/L) = 1{,}170\ lb/d$$

These daily rates are then converted to annual rates:

$$Annual\ Cl_2\ dosage\ rate = (daily\ Cl_2\ rate)\ (d/yr)$$
$$= (3{,}500\ lb/d)\ (365\ d/yr) = 1{,}278{,}000\ lb/yr$$

$$Annual\ SO_2\ dosage\ rate = (daily\ SO_2\ rate)\ (d/yr)$$
$$= (1{,}170\ lb/d)\ (365\ d/yr) = 427{,}000\ lb/yr$$

The annual costs for Cl_2 and SO_2 addition are calculated as follows:

$$Annual\ Cl_2\ cost = (annual\ Cl_2\ dosage\ rate)\ (Cl_2\ cost\ in\ \$/lb)$$
$$= (1{,}278{,}000\ lb/yr)\ (\$1.00/lb) = \$1{,}278{,}000/yr$$

$$Annual\ SO_2\ cost = (annual\ SO_2\ dosage\ rate)\ (SO_2\ cost\ in\ \$/lb)$$
$$= (427{,}000\ lb/yr)\ (\$1.20/lb) = \$512{,}000/yr$$

17. *Water Quality Issues Solution 17.* [dj] (continued)

Total operating cost = Annual Cl_2 cost + Annual SO_2 cost
= \$1,278,000/yr + \$512,000/yr = \$1,790,000/yr

The straight-line depreciation of the UV system is then calculated as follows:

Depreciation = (initial value - salvage value)/(service life)
= (\$12,000,000 - \$0)/(20 yr) = \$600,000 /yr

The total annual cost of the UV system is then calculated as:

Total annual cost = (straight-line depreciation) + (operating cost)
= (\$600,000/yr) + (\$500,000 /yr) = \$1,100,000/yr

The annual cost savings for switching to the UV system is determined as follows:

Annual cost savings = (Scenario A operating cost) - (Scenario B operating cost)

where Scenario A is the total annual Cl_2 cost, and Scenario B is the total annual UV cost.

Annual cost savings = (Annual Cl_2 cost) - (Annual UV cost)
= (\$1,790,000/yr) - (\$1,100,000/yr) = \$690,000/yr

The expected ROI on the investment in the UV system is determined as follows:

ROI = 100 (Annual cost savings)/(investment)
= 100 (\$690,000/yr)/(\$12,000,000) = 100 (0.0575) = 5.8%

This is less than the 8.0% simple interest the utility could make from cash in its bank account. Based on operating cost considerations alone, the utility should not invest in the UV system.

However, other considerations may make UV more attractive. There may be hidden costs to continued Cl_2 and SO_2 usage, such as maintaining safety equipment and training, hazardous materials planning, and, perhaps, capital costs to maintain or upgrade the Cl_2 and SO_2 storage and dosing systems. Additional calculations should be made that include these hidden costs before a final decision is made.

18. *Water Quality Issues Solution 18.* [sl]. The equilibrium partitioning equation can be linearized by taking the log of both sides of the equation. This yields the following equation:

$$\log(q) = \log(K) + n \log(c)$$

A plot of $\log(q)$ versus $\log(c)$ yields a straight line if this relationship can be used to represent the experimental data. The slope of this line is equal to n, while the intercept is $\log(K)$. The experimental data analyzed using the equation above are plotted in Figure 11 below, showing that the data fit this linearized isotherm quite well.

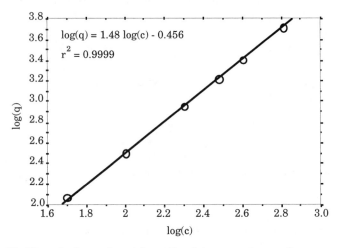

Figure 11. Linearized experimental sorption data regression results.

The regression equation presented in Figure 11 indicates that n = 1.48, and $\log(K)$ = -0.456, or K = 0.35.

19. *Water Quality Issues Solution 19.* [car]. Some of the water quality issues that affect the operation of water treatment plants treating domestic water supplies include:

- Meeting drinking water standards and regulations.
- Performing the required monitoring and analyses.
- Disposal of generated treatment plant wastes.
- Maintenance of the water treatment plant and distribution system.

20. *Water Quality Issues Solution 20.* [car]. The three major process components used to treat water sources for domestic potable water supplies are:

20. *Water Quality Issues Solution 20.* [car] (continued)

- Chemical coagulation/flocculation and softening which are utilized to remove turbidity and hardness in the raw water.
- Sedimentation and filtration which are used to remove flocculated colloidal material and precipitated hardness (calcium and magnesium) species, and possibly precipitated heavy metals.
- Disinfection which is used to reduce pathogenic organisms to acceptable levels and to provide disinfection protection in the form of a chemical residual within the water distribution system.

21. *Water Quality Issues Solution 21.* [car]. Some of the important steps that should be included in the Waste Load Allocation (WLA) process include:

- Identify the designated water use or uses for the receiving water body such as recreation, domestic water supply, agriculture, etc., and the requisite water quality standards which are enforceable by local, state or interstate agencies which promulgate them.
- An analysis of the cause-and-effect relationship between present and projected waste load inputs and subsequent water quality response through the use of:
  - site-specific field data or data from related areas and a calibrated and verified mathematical model.
  - a simplified modeling analysis based on the literature, other studies and engineering judgment.
- A sensitivity and projection analysis for achieving water quality standards under various waste load inputs.
- Determination of the "factor of safety" to be employed through, for example, a set-aside of reserve waste load capacity.
- With the residual load accounted for, an evaluation of:
  - the individual costs to the dischargers.
  - the regional cost to achieve the load and concomitant benefits of improved water quality.
- The feasibility of including discharge trading for specific contaminants.
- Development of a final waste load allocated to each discharger.

22. *Water Quality Issues Solution 22.* [car].

    a. NPDES stands for the National Pollutant Discharge Elimination System, a permit program established for each point source discharge in the United States under the Clean Water Act of 1972.

    b. The three major parameters regulated by the NPDES permit program for municipal wastewater treatment plants and their regulated maximum concentrations are shown in Table 21:

Table 21
Effluent BOD₅ and TSS Concentrations for a Municipal Wastewater Treatment Plant as Defined by the NPDES Permit Program

	30-day Average Concentration	7-day Average Concentration
5-day BOD (BOD_5)	30 mg/L	45 mg/L
TSS	30 mg/L	45 mg/L
pH	6 to 9	6 to 9

23. *Water Quality Issues Solution 23.* [car]. The BOD_5 7-day average concentration based on the data tabulated in the problem statement is 44.9 mg/L, while the 7-day average concentration for TSS is 44.7 mg/L. The wastewater treatment plant is still within its NPDES permit limit of an average 7-day maximum concentration of 45 mg/L for both BOD_5 and TSS.

24. *Water Quality Issues Solution 24.* [car]. Some of the items that should be evaluated when developing water quality objectives for a water resource include:

    a. The intended use of the water resource, i.e.,
- Domestic potable supply versus non-potable supply
- A specific industrial process
- Recreational and/or aesthetic quality uses
- Agricultural and/or sivicultural uses
- Wetland uses

    b. Consumptive versus non-consumptive uses

    c. Values to be protected
- Human health
- Aquatic ecology
- Economy and development
- Cultural values

25. *Water Quality Issues Solution 25.* [dks]. The pore water velocity, v = 5 x 10⁻³ cm/s, and the retardation factor, R = 70, are given in the problem statement. Thus, the velocity of the naphthalene is determined as follows:

$$v_c = \frac{v}{R} = \frac{5 \times 10^{-3} \text{ cm/s}}{70} = 7.14 \times 10^{-5} \text{ cm/s}$$

The naphthalene travel time is simply the distance divided by naphthalene's retarded velocity:

$$t = \frac{1,000 \text{ m}}{7.14 \times 10^{-5} \text{ cm/s } (1 \text{ m}/100 \text{ cm})} = 1.4 \times 10^{9} \text{ s} = 44.4 \text{ yr}$$

This long travel time is due to naphthalene's extremely low retarded velocity.

26. *Water Quality Issues Solution 26.* [dks].

   a. The possibility exists for recovering some of the benzene, but the success of this recovery effort depends on how far the benzene has spread and how much of the benzene has volatilized. Benzene's density is less than that of water (it is an LNAPL) so it will float on top of the groundwater as it dissolves. This fact allows contact of the free product with the air in the soil pores above the water table, and since benzene has a high vapor pressure (≈ 0.1 atm at 20°C), over time, a significant portion of the residual benzene will volatilize into the unsaturated zone of the soil. In this case, some of the benzene may be recovered using soil vapor extraction. Once the free product phase is gone, volatilization will be reduced due to benzene's high water solubility and resulting low Henry's (air to water distribution) constant. Free product may be recovered from the groundwater table by pumping if product recovery is initiated before the product has dissipated by volatilization or dissolution; otherwise, only the dissolved plume remains and significant mass recovery through pumping and treating the contaminated groundwater would not be expected to be highly effective.

26. *Water Quality Issues Solution 26.* [dks] (continued)

    b. The maximum concentration expected in the groundwater in the presence of significant amounts of free product benzene would be approximately equal the aqueous solubility of benzene, or approximately 1,780 mg/L. After the source is removed, the concentration would decline due to the dilution by groundwater flow and by natural attenuation due to biodegradation.

    c. The movement of benzene would be retarded by the presence of the organic matter in the aquifer solids. The retardation factor is determined using the formula and benzene and aquifer input data given in the problem statement. Thus, an estimate of the retardation factor is:

$$R = 1 + \frac{(0.005 \text{ g C/g soil}) (100 \text{ mL/g}) (2 \text{ g/mL})}{(0.30)} = 1 + 3.33 = 4.33$$

or the benzene will travel less than 1/4 the speed of the aquifer pore water velocity.

    d. Among aromatic compounds, benzene has a relatively high biodegradability. In fact, studies have shown that in many aquifer systems, benzene will degrade in the presence of excess oxygen at rates ranging from 0.1 to 10%/d at temperatures of 10°C, as long as the concentrations are below levels that would inhibit growth of microorganisms in the subsurface. This is an additional benefit in light of the significant retardation factor for benzene in these aquifer solids.

27. *Water Quality Solution 27.* [dks]. First, the system is sketched as indicated in Figure 12, showing the section of Echo Creek affected by this uncontrolled animal waste release in its riparian zone.

    a. The total inflow from the ranch is determined based on the continuity equation as indicated below:

$$Q = V A = V L \, d_r = (12 \text{ m/d}) (1,000 \text{ m}) (0.5 \text{ cm}) (1 \text{ m/100 cm})$$
$$= 60 \text{ m}^3/\text{d (two sides of the stream)} = 120 \text{ m}^3/\text{d}$$

27. *Water Quality Issues Solution 27.* [dks] (continued)

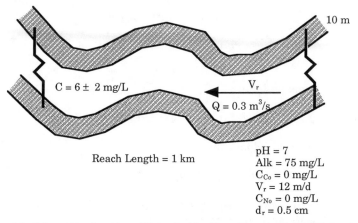

pH = 7
Alk = 75 mg/L
C_{Co} = 0 mg/L
V_r = 12 m/d
C_{No} = 0 mg/L
d_r = 0.5 cm

Reach Length = 1 km

Figure 12. Schematic of section of Echo Creek affected by an uncontrolled animal waste release.

The nitrogen and carbon concentrations in the river flow are then estimated based on the assumptions that the stream can be modeled as a plug flow reactor, that no depletion of nitrogen or carbon takes place over time, and that the system is operating at steady-state. Based on these assumptions, a mass balance equation can be written as follows:

$$\frac{dC}{dx} = k_1 (C_o - C)$$

with boundary conditions C = 0 and N = 0 at x = 0. Solving the integration yields:

$$C = C_o \left(1 - e^{-\frac{k_r x}{V_r}}\right)$$

Substitution yields the following results:

$$C = C_o \left(1 - e^{-\frac{(0.3\ d^{-1})(10\ m)}{(12\ m/d)}}\right) = C_o (1 - e^{-0.25}) = 0.22\ C_o$$

Using this equation the concentrations for nitrogen and carbon in the river inflow are:

$$C = 0.22\ (2{,}000\ g/m^3) = 442\ g/m^3$$

27. *Water Quality Issues Solution 27.* [dks] (continued)

$$N = 0.22 \, (10,000 \text{ g/m}^3) = 2,200 \text{ g/m}^3$$

From these concentrations the total loadings of carbon and nitrogen into the stream are:

Carbon loading:

$$L_C = Q \, C = (120 \text{ m}^3/\text{d}) \, (442 \text{ g/m}^3) = 53,040 \text{ g/d}$$

Nitrogen loading:

$$L_N = Q \, C = (120 \text{ m}^3/\text{d}) \, (2,200 \text{ g/m}^3) = 264,000 \text{ g/d}$$

The total in-stream concentrations are determined from a combination of the natural stream flow, $Q_s = 0.3 \text{ m}^3/\text{s} = 25,920 \text{ m}^3/\text{d}$, and the runoff flow.

$$Q_T = Q_s + Q_r = 25,920 + 120 = 26,040 \text{ m}^3/\text{d}$$

The stream carbon and nitrogen concentrations downstream of the impacted reach are determined from a mass balance based on the total loadings and the total flow in the stream after runoff:

$$C_s = (L_C)/Q_T = (53,040 \text{ g/d})/(26,040 \text{ m}^3/\text{d}) = 2.04 \text{ g/m}^3$$

$$N_s = (L_N)/Q_T = (264,000 \text{ g/d})/(26,040 \text{ m}^3/\text{d}) = 10.1 \text{ g/m}^3$$

The predicted downstream concentration of organic matter is $2.04 \text{ g/m}^3 = 2.04 \text{ mg/L}$. The measured concentration from the problem statement is 6 ± 2 mg/L ($\pm$ std. error with n = 10 observations). The question is: is 2.04 significantly different from 6 ± 2?
A t-test is used to determine whether 6 is different from 2.04 with these statistics: $t = (6 - 2.04)/2 = 1.98$. From t-tables in any statistics book it is determined that a value of 1.98 with $10 - 1 = 9$ degrees of freedom is significant at the 93% confidence level. So, at a confidence level of 95%, the two values are not different. Therefore, the existence of other sources of organic matter into the stream is not supported.

27. *Water Quality Issues Solution 27.* [dks] (continued)

    b. The unionized ammonia concentration calculation is somewhat more complex, requiring both the downstream total ammonia concentration and the pH. First, the known quantities of total nitrogen and alkalinity are converted to molar concentrations as follows:

$$N_s = 10.1 \text{ g/m}^3 \, (1 \text{ gmol/17 g}) = 0.60 \text{ gmol/m}^3$$
$$= 0.0006 \text{ gmol/L} = 6 \times 10^{-4} \text{ M}$$

$$Alk_{us} = 75 \text{ mg/L as CaCO}_3 \, (1 \text{ eq/50,000 mg})$$
$$= 0.0015 \text{ eq/L} = 0.0015 \text{ N}$$

Since the upstream pH = 7, all of the alkalinity can be assumed to be in the form of HCO_3^- which has 1 equivalent/gmol. Therefore:

$$[HCO_3^-] = 0.001498 \text{ M} \approx 0.0015 \text{ M}.$$

In the runoff, alkalinity is primarily due to the leaching of ammonia from surface soils. Assuming all NH_3 dissolved in the runoff is in the NH_3(aq) form, the contribution of run-off to the alkalinity is (from above) 6.0×10^{-4} M = 6.0×10^{-4} N = 30 mg/L as $CaCO_3$. To determine the NH_3 concentration in the stream, the pH of the stream must also be determined. If it is assumed that the system is open and in equilibrium with atmospheric CO_2, then:

$$[H_2CO_3] = (PCO_2) \, (K_H) = 10^{-5} \text{ M}$$

This means that the total amount of carbonate species in the stream depends on the pH.
A mass balance on nitrogen and carbonate species in the stream can be written as:

$$N_T = 6.0 \times 10^{-4} \text{ M} = [NH_3 \, (aq)] + [NH_4^+]$$
$$C_T = ? = [H_2CO_3^*] + [HCO_3^-] + [CO_3^=]$$

The chemical equilibria are as follows:

$$(H^+) \, (OH^-) = 10^{-14} \qquad \frac{(H^+) \, (HCO_3^-)}{(H_2CO_3^*)} = 10^{-6.3}$$

27. *Water Quality Issues Solution 27.* [dks] (continued)

$$\frac{(\overset{+}{H})(NH_3)}{(NH_4^+)} = 10^{-9.3} \qquad \frac{(\overset{+}{H})(CO_3^=)}{(HCO_3^-)} = 10^{-10.3}$$

A charge balance to determine the pH is then written knowing that the water must be electrically neutral, and that all of the positive charges must balance the negative charges. This results in the following charge balance equation:

$$[H^+] + [NH_4^+] + Z = [OH^-] + [HCO_3^-] + 2\,[CO_3^=]$$

in which Z is the concentrations of charges from cations associated with the upstream alkalinity contribution to the stream. These equations are then solved for the $[H^+] = 2.63 \times 10^{-9}\ M = 10^{-8.58}$, yielding a pH = 8.58.

The total nitrogen, mass balance and equilibrium relations presented above can now be used to determine the unionized ammonia concentration at the downstream end of the ranch as follows:

$$N_T = 6.0 \times 10^{-4} = [NH_3\,(aq)] + [NH_4^+]$$

$$= [NH_3\,(aq)] + \left(\frac{(\overset{+}{H})(NH_3\,(aq))}{10^{-9.3}}\right) = [NH_3\,(aq)] + \left(1 + \frac{(\overset{+}{H})}{10^{-9.3}}\right)$$

$$= [NH_3\,(aq)] + \left(1 + \frac{10^{-8.58}}{10^{-9.3}}\right) = [NH_3\,(aq)] + (1 + 5.25)$$

$$N_T = 6.25\,[NH_3\,(aq)]$$

which results in an aqueous ammonia concentration of:

$$[NH_3\,(aq)] = \frac{N_T}{6.25} = \frac{6.0 \times 10^{-4}\ M}{6.25} = 0.96 \times 10^{-4}\ M\ (17{,}000\ mg/gmol)$$

$$= 1.63\ mg/L$$

Chapter 8

SOLID WASTE MANAGEMENT ISSUES

Monica Minton and Poa-Chiang Yuan

I. INTRODUCTION

For the past several decades, disposal of the solid waste produced by our society, including industries and homes, has drawn a great deal of public attention. The term solid waste includes all of the heterogeneous mass of throw-aways from urban areas, as well as the more homogeneous accumulation of agricultural, industrial and mineral wastes. In urban communities the accumulation of solid waste is a direct and primary consequence of life. This accumulation adds up to approximately 4 pounds of solid waste per person every day in the U.S.

When the population was smaller, the available landfill space seemingly unlimited, the waste materials predominantly natural (as opposed to synthetic), and the understanding of possible harmful effects minimal, public concern regarding solid waste management was usually limited to quite local issues of dependable collection and mitigation of nuisance factors such as truck traffic and odor. However, times have changed. The population has increased, the use of synthetic materials is widespread, and many regions in the U.S. have only limited landfill space remaining. Advances have been made in our understanding of the chemical and physical properties of waste materials and the changes they undergo in the natural environment. Public acceptance for recycling has increased and their tolerance for risks has decreased.

The need for a nationally based solid waste management system became apparent in the 1970s. Properly dealing with large amounts of complex waste is what solid waste management is all about. In 1976, the Resource Conservation and Recovery Act (RCRA) was enacted to deal with the problems of solid and hazardous waste collection, transport, treatment, storage, and disposal in the U.S. The goals of RCRA are:

1. to protect human health and the environment;
2. to reduce waste and conserve energy and natural resources; and
3. to reduce or eliminate the generation of hazardous waste as expeditiously as possible.

Subtitle D of RCRA applies to municipal solid waste facilities, and defines municipal solid waste as follows:

- Garbage: milk cartons, coffee grounds, etc.
- Refuse: metal scraps, wallboard, etc.
- Sludge: from a waste treatment plant or a pollution control facility
- Other discarded materials, including solids, semi-solids, liquids, or contained gaseous materials resulting from industrial, commercial, mining, agricultural, and domestic sources, e.g., boiler slag, fly ash, etc., that is not otherwise considered hazardous.

Subtitle C applies to hazardous waste management, and requires all generators, transporters, storers, treaters, and disposers of hazardous waste to comply with a complex set of administrative requirements for record keeping, conformance to manifest systems, development of plans and training for securing facility permits, and to provide financial assurances to cover closure costs for their facilities.

The predominant method for municipal and hazardous waste disposal in the U.S. remains landfilling despite the emphasis that RCRA has on waste minimization and recycling. Current U.S. EPA initiatives on pollution prevention (see Chapter 4) have encouraged the analysis of alternatives to landfilling of solid waste. Recently, these alternative and combined methods for solid waste management have received considerable attention, achieved some success and simultaneously introduced other concerns. For example, incineration of solid waste greatly reduces the need for landfill space and, if so designed, can recover some of the energy value of the waste being treated. However, incineration can produce harmful air quality effects and generates an ash which concentrates heavy metals that may require secure landfill disposal. Composting of organic materials is another alternative to landfilling that reduces volume and produces a useful soil conditioner. It, however, requires significant space and operator attention, and can generate undesirable odor problems.

In considering these issues of municipal and hazardous waste management this chapter provides basic problems and calculations relevant to the description and evaluation of a variety of waste management practices. These practices include conventional sanitary landfilling as well as alternative management approaches including composting, recycling, and incineration/waste to energy.

II. PROBLEMS

1. *Solid Waste Management Issues Problem 1.* (municipal solid waste characteristics). [sn]. Besides economic considerations and site specifications, what are the major characteristics of municipal solid waste that are employed for the selection of the treatment and disposal processes listed below?

    a. Composting.
    b. Incineration.
    c. Sanitary landfill.

2. *Solid Waste Management Issues Problem 2.* (basic principles). [sn]. The physical state of hazardous material in a storage or transport container is an important factor in considering the fate of transport into the environment during an accidental spill or discharge situation. Define the conditions of container temperature (T_c) and ambient temperature (T_a) that will maintain the status of a chemical with the following physical characteristics. Note that MP and BP are the melting point and boiling point of the chemical, respectively.

Problem	Physical State of Material	MP/BP	Container Conditions
Example	Cold or Refrigerated Solid	$MP > T_a$	$T_c < MP$ & T_a
a.	Solid		
b.	Warm or Hot Liquid		
c.	Cold Liquid		
d.	Liquid		
e.	Hot Liquid		
f.	Hot or Warm Compressed Gas or Vapor Over Hot Liquid		
g.	Compressed Liquefied Gas		
h.	Hot or Warm Compressed Gas or Compressed Liquefied Gas		

3. *Solid Waste Management Issues Problem 3.* (municipal solid waste management, quantity, measurement). [sxl]. What are the most common methods used to estimate the quantity of waste generated in a community?

4. *Solid Waste Management Issues Problem 4.* (municipal solid waste management, waste measurement). [sxl]. A municipality in the Midwest has a population of 50,000 and generates 100,000 yd^3 of municipal waste annually. The waste is

4. *Solid Waste Management Issues Problem 4.* [sxl] (continued)

made up of 30% compacted waste and 70% uncompacted waste. Assume that the waste has a density of 1,000 lb/yd³ compacted, and 400 lb/yd³ uncompacted. How many pounds of waste are generated by this city each year? By each person each year?

5. *Solid Waste Management Issues Problem 5.* (municipal solid waste management, waste quantity, measurement). [sxl]. An analysis of a solid waste generator has revealed that the waste is composed (by volume) of 20% supermarket waste, 15% plastic-coated paper waste, 10% polystyrene, 20% wood, 10% vegetable food waste, 10% rubber, and 10% hospital waste. What is the average density of this solid waste in lb/yd³? Use the following as discarded densities (lb/yd³) for each of the components of the generator's waste: supermarket waste, 100; plastic-coated paper waste, 135; polystyrene, 175; wood, 300; vegetable food waste, 375; rubber-synthetics, 1,200; and hospital waste, 100.

6. *Solid Waste Management Issues Problem 6.* (municipal solid waste, moisture content). [pcy]. The design of garbage collection vehicles differs from country to country based on location, local culture, etc. Many factors influence the design of these vehicles, with moisture content being one of the most important considerations in the hauling of municipal waste. The general formula for calculating the moisture content of solid waste is as follows:

$$\text{Moisture Content} = \frac{(A - B)}{A}(100)$$

where A = weight of sample as delivered, kg; and B = weight of sample after drying, kg.

Use the data provided below on waste component moisture content and the weight composition of Jefferson County, Mississippi waste to determine the average moisture content of their municipal solid waste. Base the calculation on a 100 kg sample.

Table 22
Typical Moisture Content of Municipal Solid Waste in the U.S.

Component	Moisture Content (wt%)
Food waste	70
Paper	6
Garden waste	60
Plastic	2
Textiles	10
Wood	20
Glass	2
Metals	3

6. *Solid Waste Management Issues Problem 6.* [pcy] (continued)

Table 23
Composition of Jefferson County, Mississippi Solid Waste on a Weight Percent Basis

Component	Composition (wt%)
Food waste	20
Paper	22
Garden waste	18
Plastic	3
Textiles	7
Wood	25
Glass	2
Metals	3

7. *Solid Waste Management Issues Problem 7.* (municipal solid waste, compaction factor). [pcy]. The Jefferson County Solid Waste Management Corporation analysis of its solid waste includes the following major components: weight, volume, and compaction factors. Determine the density of well-compacted waste as delivered to a landfill. (Note: waste components and compaction factors vary from place to place, and may not be identical to those shown below for Jefferson County.)

Table 24
Composition, Weight, Volume and Compaction Factor of Solid Waste from Jefferson County, Mississippi

Component	Weight, kg	Volume as Discarded (m³)	Compaction Factor
Food waste	250	1.00	0.33
Paper	300	3.75	0.15
Garden waste	250	1.18	0.20
Plastic	50	0.80	0.10
Textiles	60	9.61	0.15
Wood	50	0.34	0.30
Glass	20	0.10	0.40
Metals	20	0.10	0.30
Total	1,000		

8. *Solid Waste Management Issues Problem 8.* (hazardous waste incineration, combustion). [sxl]. In many cases, hazardous waste can be converted to inorganic matter or changed in form by high-temperature processes in the presence of oxidizing agents. Such thermal processes can result in the partial or complete reduction in the degree of hazard of a material. The general class of such thermal processes when oxygen is the oxidizing agent is termed incineration. What conditions must be present in an incinerator to provide successful conversion of a hazardous waste material?

9. *Solid Waste Management Problem 9*. (hazardous waste, hazardous waste incinerator, destruction and removal efficiency, trichloroethylene, POHCs). [jr]. By federal law, hazardous waste incinerators must meet a minimum destruction and removal requirement, i.e., that the principal organic hazardous constituents (POHCs) of the waste feed be incinerated with a minimum destruction and removal efficiency (DRE) of 99.99% ("four nines"). The destruction and removal efficiency is the fraction of the inlet mass flow rate of a particular chemical that is destroyed and removed in the incinerator or, equivalently,

$$DRE = 1 - \frac{\text{mass flow of chemical out}}{\text{mass flow of chemical in}}$$

A feed stream to an incinerator contains 245 kg/hr of trichloroethylene. Calculate the maximum outlet flow rate of trichloroethylene from the incinerator allowed by law.

10. *Solid Waste Management Issues Problem 10*. (hazardous waste, hazardous waste incinerator, particulate removal). [jr]. Current federal regulations limit stack emissions of particulate material from a hazardous waste incinerator to an outlet loading of 0.08 gr/dscf (grains of particulate per dry standard cubic foot of stack gas), corrected to 7% oxygen (or an approximate excess air level of 50%).

　　a.　Describe the effect on the outlet particulate loading of each of the three following actions as far as increasing or decreasing the measured value of the outlet loading. Give a reason for each answer.
　　　　• Removing water from the gas. (Note: The incineration of organic material almost always produces water as a product.)
　　　　• Lowering the temperature of the gas from its stack temperature of 800°F to the standard temperature of 60°F.
　　　　• Using 50% excess air for the incineration instead of 100% excess air.
　　b.　From your answers to Part a, explain why the terms "dry," "standard," and "corrected to 7% oxygen" are used in the regulations.

11. *Solid Waste Management Issues Problem 11*. (hazardous waste incinerator, particulate removal). [jr]. An incinerator is burning a hazardous waste using 50% excess air. The stack gas at 800°F has a flow rate of 10,000 acfm (actual cubic feet per minute) with a particulate mass flow of 20 gr/min and a water content of

11. *Solid Waste Management Issues Problem 11.* [jr] (continued)

10% by volume. The maximum allowed outlet loading for particulates is 0.08 gr/dscf (grains per dry standard cubic foot), corrected to 7% oxygen (or an approximate excess air level of 50%).

    a. Calculate the outlet loading in gr/acf (grains per actual cubic foot).
    b. Calculate the outlet loading in gr/scf (grains per standard cubic foot) noting that standard temperature is 60°F.
    c. Calculate the outlet loading in gr/dscf.
    d. As far as particulates are concerned, is the incinerator in compliance?

12. *Solid Waste Management Issues Problem 12.* (oil spill, risk). [sn]. There have been a number major accidents involving oil spills in the U.S. such as those occurring near Prince William Sound, Alaska (Exxon Valdez); Galveston Bay, Texas; and Lake Michigan, Michigan. These incidents resulted in extensive environmental damage and massive cleanup efforts by the responsible parties, and state and federal agencies. What are the possible risks of these incidents to ecological systems and to humans? Briefly describe the mechanisms of the risk (if any).

13. *Solid Waste Management Issues Problem 13.* (municipal solid waste, composting). [sn]. Composting is the biological decomposition of organic waste material by microorganisms. It is one of the major treatment alternatives for municipal solid waste, especially in developing countries where solid wastes contain mostly organic carbon sources and are rich in nutrients. During the process, certain physical, chemical and biological changes take place which alter the character of the waste material. Describe these physical, chemical, and biological changes as municipal solid waste is converted to humus in the composting process.

14. *Solid Waste Management Issues Problem 14.* (industrial solid waste management, hazardous waste management, Superfund, Love Canal). [sn]. The Love Canal is a hazardous site located in the industrial community of Niagara Falls, New York. It is the most notorious waste site in U.S. history, having led to the passage of the Superfund legislation in 1980. It symbolizes a tragedy that finally compelled assertive government action to correct decades of waste disposal abuses. The incident has had a significant impact on current hazardous waste management practices.

    Since the Love Canal incident, what changes have been observed in the attitude of government, industry and the general

14. *Solid Waste Management Issues Problem 14.* [sn] (continued)

public toward management of industrial solid and hazardous waste?

15. *Solid Waste Management Issues Problem 15.* (solid waste management, recycling). [fwk]. Choose a small political jurisdiction that operates its own solid waste collection system. Find out the quantity of solid waste collected, any waste processing that takes place, and the ultimate disposal option used for this waste stream. Obtain cost information for waste collection and waste disposal.

Find out what type of recycling, if any, is done in the jurisdiction. For a class of materials not yet widely recycled, develop a plan to collect, process, and sell this recycled product. Make rough engineering estimates of the cost of operating the recycling process. Determine the savings incurred because of reduced waste collection and/or disposal costs. As recycling is almost always not economically attractive, look for other values to society that this recycling represents, and incorporate these ideas as quantitatively as possible into a persuasive report selling your project.

III. SOLUTIONS

1. *Solid Waste Management Issues Solution* 1. [sn]. The characteristics relevant to the treatment and disposal options listed in the problem statement are summarized below:

   a. Composting - Biodegradable/non-biodegradable fraction, moisture content, carbon to nitrogen ratio.
   b. Incineration - Heating value, moisture content, inorganic fraction, metal content, alkali earth metal content.
   c. Sanitary landfill - Biodegradable to non-biodegradable fraction, liquid content, hazardous material content.

2. *Solid Waste Management Issues Solution* 2. [sn]. The following is a summary of the solution to this problem.

Problem	Physical State State of Material	MP/BP	Container Conditions
Example	Cold or Refrigerated Solid	$MP > T_a$	$T_c < MP$ & T_a
a.	Solid	$MP > T_a$	T_c near T_a
b.	Warm or Hot Liquid	$BP > T_a$	$T_c > MP$ & $T_a < BP$
c.	Cold Liquid	$MP < T_a$	$T_a > MP < T_a$ & BP
d.	Liquid	$BP > T_a$	T_c near T_a
e.	Hot Liquid	$BP > T_a$	$BP > T_c > T_a$
f.	Hot or Warm Compressed Gas or Vapor Over Hot Liquid	$BP > T_a$	$T_c > BP$ & T_a
g.	Compressed Liquefied Gas	$BP < T_a$	T_c near T_a
h.	Hot or Warm Compressed Gas or Compressed Liquefied Gas	$BP < T_a$	$T_c > BP$ & T_a

3. *Solid Waste Management Issues Solution 3*. [sxl]. The first and most accurate method to estimate the quantity of solid waste generated is to weigh the waste requiring disposal for a defined period of time and normalizing this result to the population from which the waste is generated. This quantity is typically expressed as lb waste generated/person/d. This total per capita quantity can then be multiplied by any projected increase in population to determine the anticipated generation rate for future years.

The second method is to determine the volume of waste generated and use known density factors to convert this measurement into an associated weight.

3. *Solid Waste Management Issues Solution 3.* [sxl] (continued)

The third and least accurate method for determining waste generation rates in a region is to determine the population of the area and multiply this by typical reported-per-capita waste generation rates to yield a projected total generation rate for the community of interest.

4. *Solid Waste Management Issues Solution 4.* [sxl]. Based on the waste densities given in the problem statement, the following generation rates are determined:

$$\text{Waste Generated/yr} = (0.3)\,(100{,}000\ \text{yd}^3)\,(1{,}000\ \text{lb/yd}^3)$$
$$+ (0.7)\,(100{,}000\ \text{yd}^3)\,(400\ \text{lb/yd}^3)$$
$$= 30{,}000{,}000\ \text{lb} + 28{,}000{,}000\ \text{lb} = 58{,}000{,}000\ \text{lb/yr}$$

$$\text{Per Capita Generation Rate} = (58{,}000{,}000\ \text{lb/yr})/50{,}000\ \text{people}$$
$$= 1{,}160\ \text{lb/person/yr} = 3.2\ \text{lb/person/d}$$

5. *Solid Waste Management Issues Solution 5.* [sxl]. Based on the volume percent composition and waste densities given in the problem statement, the following average as discarded waste density is estimated to be:

$$\text{Average Density} = (0.2)\,(100) + (0.15)\,(135) + (0.15)\,(175)$$
$$+ (0.2)\,(300) + (0.1)\,(375) + (0.1)\,(1200) + (0.1)\,(100)$$
$$= 294\ \text{lb/yd}^3$$

6. *Solid Waste Management Issues Solution 6.* [pcy]. The dry weight of each component can be determined using the following equation:

$$\text{Dry Weight} = (\text{Discarded Weight})\,(1 - \%\ \text{Moisture}/100)$$

The Discarded Weight for each component can be determined using the following equation:

$$\text{Discarded Weight} = (\text{Total Waste Weight})\,(\text{Weight}\ \%\ \text{of Component})$$

Based on the % moisture content data and waste composition data provided in the problem statement in Tables 22 and 23, respectively, and assuming a Total Waste Weight of 100 kg, the calculations presented in Table 25 result for each waste component given in the problem statement.

Finally, the average moisture content of the waste is determined as:

$$\text{Average}\ \%\ \text{Moisture} = [(\text{Discarded Weight} - \text{Dry Weight})/100]\,(100)$$

6. *Solid Waste Management Issues Solution 6.* [pcy] (continued)

Table 25
Composition, Weight Percent, Component Discarded Weight, and
Component Dry Weight for the Municipal Solid Waste Stream from
Jefferson County, Mississippi

Component	Wt%	Component Discarded Weight (kg)	Component Dry Weight (kg)
Food waste	20	20	(1.00 - 0.7) (20) = 6.00
Paper	22	22	(1.00 - 0.6) (22) = 20.68
Garden waste	18	18	(1.00 - 0.6) (18) = 7.20
Plastic	3	3	(1.00 - 0.02) (3) = 2.91
Textiles	7	7	(1.00 - 0.10) (7) = 6.30
Wood	25	25	(1.00 - 0.20) (25) = 20.00
Glass	2	2	(1.00 - 0.02) (2) = 1.96
Metals	3	3	(1.00 - 0.03) (3) = 2.91
Total	100		67.96 = 68

Average % Moisture = [(100 - 68)/100] (100) = 32%

7. *Solid Waste Management Issues Solution 7.* [pcy]. The compacted volume of each component in the landfill can be determined using the following equation:

Compacted Volume = (Discarded Volume) (Compaction Factor)

Based on the weight, discarded volume, and compaction factor data provided in the problem statement in Table 24, the calculations presented in Table 26 result for each waste component given in the problem statement.

Finally, the average density of the well-compacted mixed waste delivered to the landfill can be determined as:

Average Density = (Discarded Weight)/(Total Compacted Volume)

Table 26
Composition, Weight, Volume, Compaction Factor, and Compacted Volume
of Solid Waste from Jefferson County, Mississippi

Component	Weight, kg	Volume as Discarded (m³)	Compaction Factor	Compacted Volume (m³)
Food waste	250	1.00	0.33	0.33
Paper	300	3.75	0.15	0.56
Garden waste	250	1.18	0.20	0.24
Plastic	50	0.80	0.10	0.08
Textiles	60	9.61	0.15	1.44
Wood	50	0.34	0.30	0.10
Glass	20	0.10	0.40	0.004
Metals	20	0.10	0.30	0.018
Total	1,000			2.77

7. *Solid Waste Management Issues Solution 7.* [pcy] (continued)

$$\text{Average Waste Density} = 1000 \text{ kg/2.77 m}^3$$
$$= 360.75 \text{ kg/m}^3$$

8. *Solid Waste Management Issues Solution 8.* [sxl]. In order for incineration to be successful in the destruction of a hazardous waste material, the following characteristics must exist:

- Adequate free oxygen must always be available in the combustion zone.
- Sufficient turbulence must exist within the incinerator to ensure the constant mixing of waste and oxygen.
- Adequate combustion temperatures must be maintained. Exothermic combustion reactions must supply enough heat to raise the burning mixture to a sufficient temperature to destroy all of the organic components of the waste.
- The duration of exposure of the waste material to combustion temperatures must be long enough to ensure that even the slowest combustion reaction is completed. In other words, transport of the burning mixture through the high temperature region must occur for a sufficient period of time to allow reactions to go to completion.

9. *Solid Waste Management Issues Solution 9.* [jr]. Substituting data given in the problem statement into the DRE equation provides the means of determining an acceptable mass flow rate of chemical out of the incinerator as follows:

$$\text{DRE} = 1 - \frac{\text{mass flow of chemical out}}{\text{mass flow of chemical in}} = 0.9999$$

$$= 1 - \frac{\text{mass flow of chemical out}}{245 \text{ kg/hr}}$$

$$\text{mass flow of chemical out} = (1 - 0.9999)(245 \text{ kg/hr})$$
$$= (0.0001)(245 \text{ kg/hr}) = 0.0245 \text{ kg/hr}$$

10. *Solid Waste Management Issues Solution 10.* [jr].

a. Removing water from the gas decreases the gas volume without affecting the mass of particulate matter. Since the particulate loading is the mass of particulate matter divided by the gas volume, the particulate loading increases.

10. *Solid Waste Management Issues Solution 10.* [jr] (continued)

Lowering the temperature also decreases the gas volume without affecting the particulate mass. Therefore, the loading increases.

The less excess air that is used for the combustion, the less oxygen and nitrogen is contributed to the gas stream. The gas volume is therefore lower and the particulate loading increases.

b. "Dry" is used so that extra water cannot be added to the gas stream to artificially lower the outlet particulate loading. "Standard" is used so that the outlet particulate loading is not a function of temperature, i.e., the loading cannot be artificially lowered by raising the temperature. "Converted to 7% oxygen" is used so that the outlet particulate loading cannot be artificially lowered by using an inordinate amount of excess air.

11. *Solid Waste Management Solution 11.* [jr].

a. The outlet loading, OL, in gr/acf is given by:

$$OL_{gr/acf} = \frac{314 \text{ gr/min}}{10,000 \text{ ft}^3/\text{min}} = 0.0314 \text{ gr/acf}$$

b. From the ideal gas law, the volume is proportional to the absolute temperature. Noting that $T(°R) = T(°F) + 460$, the outlet loading in gr/scf is:

$$OL_{gr/scf} = (0.0314 \text{ gr/ft}^3)\left(\frac{800 + 460°R}{60 + 460°R}\right) = 0.0761 \text{ gr/scf}$$

c. Since the gas stream is 10% water by volume, the outlet loading in gr/dscf is given by:

$$OL_{gr/dscf} = (0.0761 \text{ gr/ft}^3)\left(\frac{100 \text{ ft}^3_{wet}}{90 \text{ ft}^3_{dry}}\right) = 0.0845 \text{ gr/dscf}$$

11. *Solid Waste Management Issues Solution 11.* [jr] (continued)

    d. Because 50% excess air (approximately 7% O_2) is used, the outlet loading is 0.0845 gr/dscf corrected to 7% O_2. This is above the allowed maximum particulate concentration of 0.08. The incinerator is not in compliance.

12. *Solid Waste Management Issues Solution 12.* [sn]. Risks to marine life and to humans from accidents involving major oil spills include the following:

Immediate effects:
- direct kill through coating and asphyxiation
- skin contact
- absorption of water soluble toxic components of oil
- destruction of the food sources of higher species (e.g., food web)

Long-term effects:
- reduce resistance to infection and stresses
- carcinogenic and potential mutagenic
- Life style interruption (e.g., propagation)

Effects to valuable commercial shellfish can result (e.g., oysters, scallops, soft-shell clams).
Risks result from the human use of marine resources (e.g., fisheries products).
Risks result from the recreational use of the marine resources from oil tar or prolonged skin contact with carcinogenic hydrocarbons that contribute to a public health hazard.
Risks result from direct water utilization.

13. *Solid Waste Management Issues Solution 13.* [sn]. Changes taking place in the physical, chemical, and biological quality of municipal solid waste during the composting process are summarized below.

Changes	Solid waste	Compost
Physical changes		
Color	Variable	Dark black
Particle size	Irregular, large	Humus-bike, small
Water content	Seasonal, Variable	Generally dryer

13. *Solid Waste Management Issues Solution 13.* [sn] (continued)

Changes	Solid waste	Compost
Chemical changes		
C/N ratio	Variable, from 20 to 78/1	$\approx 20/1$
Organic substrate	Unstable	Stable
Amount of NO_3 & NH_4	Relatively low	Increased due to degradation of proteins
Biological changes		
Microbial activity	High	Stabilized, decrease in
Number of pathogens	High	both microbial activity and pathogen due to aerobic pile conditions and heat generated during decomposition

14. *Solid Waste Management Issues Solution 14.* [sn]. The Love Canal incident occurred in the mid-1970s. Since then, hazardous waste problems have led the public to initiate action and have brought attention to government agencies, scientists, and professionals. There are quite a number of local, state, and national organizations involved with the management of hazardous waste. A successful example of a national organization which grew out of Love Canal is the "Citizens Clearing House for Hazardous Waste." This organization acts as a resource center to provide community residents with scientific information on hazardous and toxic materials including the toxic effects of chemicals, landfill technology, and alternate disposal of hazardous waste. Other services include linking people and relevant government agencies together, helping communities organize, and referring people to sources of technical and legal help.

In some cases, a community health profile of people living near hazardous waste disposal sites has been initiated to collect information about demographics and health problems. This has been found to be useful to public health professionals in planning risk assessment studies of hazardous waste sites. It was also found that people who live near hazardous waste disposal sites have more concerns about their health and educate themselves about the toxicity of chemicals that have been found on the site.

From a government perspective, after the Love Canal incident, government and state agencies have come to realize that mismanagement of hazardous waste has created substantial risk to human health and the environment. The Resource Conservation and Recovery Act (RCRA) was enacted in 1976 to deal with the problems of solid and hazardous waste. RCRA was later amended

14. *Solid Waste Management Issues Solution 14.* [sn] (continued)

in 1984 (the Hazardous and Solid Waste Amendments of 1984). Since then, hazardous wastes have been regulated based on the "cradle to grave" concept. Under these laws, the government requires existing and future treatment, storage, and disposal facilities (TSDFs) to maintain operating records and develop a groundwater monitoring program during operation and post-closure periods of disposal. Financial responsibility requirements for hazardous waste facilities were also promulgated by U.S. EPA to prevent the abandonment or improper closure of hazardous waste facilities and their attendant hazards.

The government also came to realize the need of resources for the clean-up of abandoned sites, which later led to the passage of the Comprehensive Environmental Response, Compensation and Liability Act of 1980 (CERCLA) or "Superfund." Since then the disadvantages of traditional pollution control practices have become evident. The environmental management of hazardous waste has been shifting toward a pollution prevention approach which is designed to limit the amount of waste generated in the first place, and to ensure that the wastes pose no threat to people's health or the environment. The Pollution Prevention Act of 1990 established pollution prevention as a national policy, and emphasized its role as the lead management tool within the hierarchy of waste management approaches.

Numerous state and municipal governments enacted worker Right-to-Know laws which led to the promulgation of a Federal Community Right-to-Know law entitled the Hazardous Communication Standard, 29 CFR 1910.1200. The objective of this Act is to communicate to the worker and to the community the presence and effects of hazardous chemicals in the workplace and within commercial facilities in their community.

From an industry point of view, many environmental laws have been formulated and enacted that lead to complex administrative requirements for record-keeping, conformance to manifest systems, development of plans and training programs for securing facility permits, etc. Also industry has to be fully aware of requirements for owners and operators of TSDFs which include the availability of funds to cover closure costs for the facility in an environmentally sound manner as addressed in the Superfund Amendments and Reauthorization Act (SARA) of 1986. Pollution prevention clearly has become a top priority for industries since it reduces the cost of waste management, regulatory compliance, liabilities, and other indirect short-term and long-term costs of doing business.

15. *Solid Waste Management Issues Solution 15.* [fwk]. This is an open-ended question that will be highly site-specific. The following discussion is provided to help guide the students and

15. *Solid Waste Management Issues Solution 15.* [fwk] (continued)

instructor in developing a comprehensive analysis and summary of conditions existing, and potential options available for waste management and recycling in a given community.

There are many specialized recycling efforts that might be overlooked. In addition to the commonly recycled products such as aluminum, ferrous metal, plastics, glass, newspapers, high quality office paper, food waste and yard waste, there are arguments to be made for programs to recycle tires, motor oil, automobile coolant, transmission fluid, car batteries, flashlight batteries, wood chips from clearing brush under power lines, discarded chemicals from public school and college laboratories, copper from discarded motors and wire, gold from computer chips and circuit boards, rare metals from automobile catalytic converters, etc.

There are places where unsorted recyclables are sorted into the separate categories of clear glass, green glass, ferrous metal, aluminum and each of the various plastics. Note the number embossed on each plastic part. Numbers 1, 2 and 6 currently have the strongest markets. Given that sorting can be done at various stages, a significant part of the design, from society's point of view, will be on where the sorting is to take place: in the consumer's home, at the point of drop off of the material, by workers on a conveyor line, or by automated physical or chemical processes.

An offshoot of this kind of project is to look at the quality control needs for a particular product stream. In this case, one would take a commonly recycled product and investigate the sources and the buyers of the material. The buyers would have specific quality requirements based on the end use of the recycled material. The questions to be explored are how is the quality of the stream maintained? How does one test to see if the quality is at a level that it should be? What are the consequences of contamination of the stream by various common contaminants? A classic example would be the contamination, on purpose, of used motor oil by someone who hopes to get rid of PCB-laden oil, and the consequences this possibility has on necessary controls over acquisition of waste oil, used oil chemical testing requirements and methods, finished product quality assurance, etc.

Chapter 9

INDUSTRIAL HYGIENE

Pankajam Ganesan

I. INTRODUCTION

Industrial hygiene is an area of specialization in the field of industrial health and safety in which the industrial hygienist is trained to anticipate, recognize, evaluate, and control health hazards that arise in the industrial work environment. It involves consideration of major factors such as chemical stresses, physical agents, and biological and ergonomic hazards that can cause sickness or serious discomfort to workers. Industrial hygiene deals with the occupational aspects of the field of environmental health

With an increased interest in the protection of the environment, both internal and external to an industrial setting, the number of alleged threats to human health and safety, both as a worker and as a resident in an industrial neighborhood, have multiplied.

The purpose of the Occupational Safety and Health Act is "to assure so far as possible every working man and woman in the nation safe working conditions and to preserve our human resources" (U.S. Department of Labor. All About OSHA 2056, 1991 (Revised) p.1). The major roles of OSHA include: developing and enforcing mandatory job safety and health standards; maintaining a reporting and record keeping system to monitor job-related injuries and illness; encouraging employers and employees to reduce workplace hazards and implement or improve safety and health programs; providing research in safety and health; establishing training programs; and establishing separate but dependent responsibilities and rights for employers and employees to achieve better safety and health conditions in the workplace.

The major concerns in the field of industrial hygiene include the various chemical stresses, and physical, biological, and ergonomic hazards that can cause sickness or serious discomfort to workers. Chemical stresses may include chemical dusts, liquids, gases, fumes, vapors, smoke, and mists that present dangers when they are swallowed, breathed, or make contact with the eyes or skin. Physical agents include noise, vibration, high and low temperatures and pressures, and ionizing and non-ionizing radiation. Biological hazards include any viruses, bacteria, fungi, parasites, or other living organism that can cause disease in human beings. Ergonomic hazards are related to the design and conditions of the workplace. Poorly designed workstations and/or tools such as any control that is difficult to reach or operate, any instrument dial that has poor legibility, any seat that induces poor posture or discomfort, or any obstruction of vision that contributes to an accident are all ergonomic hazards.

The industrial hygienist is largely responsible for the health and well-being of the industrial employee. Chemical, physical, biological, and ergonomic recommendations usually bring about, directly or indirectly, improved job performance with increased safety, health, and well-being of the employee. Of course, both management and employee cooperation is essential to increase the ease and efficiency at a workplace.

The improved safety and health of today's work environment are the direct result of concentrated efforts by workplace safety and health and industrial hygiene teams working with management that focus on enhancing an organization's primary asset, a safe and healthy work force. The problems presented in this section cover general personal protective equipment, noise, fire protection, ergonomics, industrial hygiene surveys, hazard communication, and mass balance concepts applied in the field of industrial hygiene.

II. PROBLEMS

1. *Industrial Hygiene Problem 1.* (unit conversions, STP, ideal gas law). [hp]. Atmospheric concentrations of toxic pollutants are usually reported using two types of units: 1) mass of pollutant per volume of air, i.e., mg/m^3, µg/m^3, ng/m^3, etc.; or 2) parts of pollutant per parts of air (by volume), i.e., ppmv, ppbv, pptv, etc. In order to compare data collected at different conditions, actual concentrations are often converted to Standard Temperature and Pressure (STP). According to EPA, STP conditions for atmospheric or ambient sampling are often: 25°C and 1 atm.

Assume the concentration of chlorine vapor is measured to be 15 mg/m^3 at a pressure of 600 mmHg and at a temperature of 10°C.

    a. Convert the concentration units to ppmv.
    b. Calculate the concentration in units of mg/m^3 at STP.

2. *Industrial Hygiene Problem 2.* (concentration, unit conversions, ideal gas law, VOCs, sorbent tubes). [hp]. To sample for volatile organic compounds (VOCs) in the atmosphere, air may be pumped through an adsorption tube during a sampling period. A solvent is then flushed through the tube to desorb the compounds from the adsorbent, and gas chromatography/mass spectrometry (GC/MS) analysis is used to identify and quantify the VOCs dissolved in the solvent.

Assume that a smoker's breath was drawn through an adsorbent tube at a rate of 498 mL/min for a period of 159 s at a temperature of 22°C and at a pressure of 624 mmHg. If the concentration of benzene in the smoker's breath was found to be 10.1 ppbv, what mass of benzene, in ng, was recovered in the solvent?

3. *Industrial Hygiene Problem 3.* (industrial hygiene, chemical protective clothing, permeation, penetration). [pcy]. What is the difference between permeation and penetration as they apply to chemical protective clothing?

4. *Industrial Hygiene Problem 4.* (industrial hygiene, chemical protective clothing). [pcy]. There are four levels of chemical protective clothing and equipment defined by the U.S. EPA. Describe each of these levels.

5. *Industrial Hygiene Problem 5.* (industrial hygiene, accident prevention). [pg]. In an industrial environment, what is the single most important factor in an accident prevention program?

5. *Industrial Hygiene Problem 5.* [pg] (continued)

Comment on your choice.

    a. A written program.
    b. Training.
    c. Safety engineers.
    d. Management support.

6. *Industrial Hygiene Problem 6.* (Threshold Limit Value, TLV). [pg]. Explain what is meant by the Threshold Limit Value or TLV.

7. *Industrial Hygiene Problem 7.* (noise). [pg]. List the temporal factors that can influence noise exposure.

8. *Industrial Hygiene Problem 8.* (noise). [pg]. Explain the mechanism of hearing and the effect of noise exposure on individuals.

9. *Industrial Hygiene Problem 9.* (industrial hygiene, stress, discomfort, chemical hazards, biological hazards, ergonomics). [pg]. Explain briefly the environmental factors or stresses that can cause sickness, impaired health or significant discomfort in workers.

10. *Industrial Hygiene Problem 10.* (fire protection, fire hazard). [pg]. Explain general factors to consider during a fire emergency.

11. *Industrial Hygiene Problem 11.* (fire protection, fire development). [pg]. Explain the four stages of a fire.

12. *Industrial Hygiene Problem 12.* (fire protection, combustion products). [pg]. List the general products of combustion during a fire and explain the hazardous exposures associated with them.

13. *Industrial Hygiene Problem 13.* (industrial hygiene management surveys, industrial hygiene plant surveys). [pg]. Explain briefly what constitutes industrial hygiene management and plant surveys.

14. *Industrial Hygiene Problem 14.* (ergonomics, anthropometry, biomechanics). [pg]. Briefly explain the meaning of ergonomics, anthropometry, and biomechanics as applied to the field of ergonomics.

15. *Industrial Hygiene Problem 15.* (ergonomics, computer workstations). [pg]. Explain the basic health problems associated with computer operators and briefly explain the basic elements to be considered in designing an office (computer) workstation.

16. *Industrial Hygiene Problem 16.* (OSHA, hazard communication standard, MSDSs). [pg]. Explain OSHA's Hazard Communication Standard.

17. *Industrial Hygiene Problem 17.* (dilution ventilation, toluene, TLV, ventilation systems, mixing factor). [kg]. Exposure to volatile organic compounds in a work place may have adverse health effects on the exposed workers based on their concentration, duration of exposure, and the toxicity of the compound. Often dilution ventilation is used to maintain a specific vapor concentration within acceptable limits.

Estimate the dilution ventilation required in a work area where a toluene-containing adhesive is used at a rate of 3 gal/8-hr work day. Assume that the specific gravity of toluene is 0.87, that the adhesive contains 40 vol% toluene, and that 100% of the toluene is evaporated into the room air at 20°C. The plant manager has specified that the concentration of toluene must not exceed 80% of its threshold limit value (TLV) of 100 ppm. The following equation can be used to estimate the dilution air requirement:

$$Q = KV/C_a$$

where Q = dilution air flow rate, ft^3/min; K= dimensionless mixing factor that accounts for less than complete mixing characteristics of the contaminant in the room, the contaminant toxicity level, and the number of potentially exposed workers. Usually, the value of K varies from 3 to 10, where 10 is used under poor mixing conditions and when the contaminant is relatively toxic (TLV < 100 ppm). For solving this problem use K = 5; V = volume of pure contaminant vapor, ft^3; and C_a = acceptable contaminant concentration in the room, ppm x 10^{-6}.

18. *Industrial Hygiene Problem 18.* (industrial hygiene, sampling, compositing, mass balance). [dks]. You have been asked to take samples and carry out a mass balance on an industrial operation that is losing a valuable material, identified only as X, to the plant drainage system. Based on available information you have deduced that the following patterns of waste flow and waste concentration are reasonable, at least for the purposes of planning your sampling program.

Based on Figures 13 and 14, below, answer the following questions:

18. *Industrial Hygiene Problem 18.* [dks] (continued)

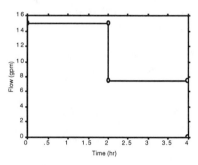

Figure 13. Flow versus time.

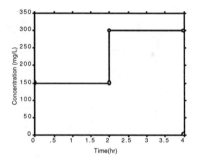

Figure 14. Concentration versus time.

    a. Calculate the average flow from Figure 13.
    b. Calculate the average concentration of X from Figure 14.
    c. Calculate the quantity of X being discharged based on your results from Parts a and b.
    d. Calculate the mass of X discharged by summing up the quantities $K (X_i) (Q_i)$, where i identifies the time interval in which $X = X_i$ and $Q = Q_i$ (i = 1, 2, 3, and 4), and K is a constant that makes the proper conversion of units.
    e. Compare the estimated quantities of X discharged according to Parts c and d, and explain why they are different.

19. *Industrial Hygiene Problem 19.* (industrial hygiene, sampling, compositing, mass balance). [dks]. You have been asked to take samples and carry out a mass balance on another industrial operation that is losing material X to its plant drainage system. You plan to use the same method that you did for Problem 18 to

19. *Industrial Hygiene Problem 19.* [dks] (continued)

determine the extent of material X loss in the system. The following patterns of waste flow and waste concentration are observed in this second plant.

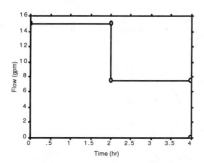

Figure 15. Flow versus time.

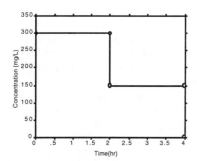

Figure 16. Concentration versus time.

    a. Use the flow and concentration pattern shown in Figures 15 and 16 above and answer Questions a through e in Problem 18 above.
    b. Make some general statements regarding the relation of flow-concentration patterns and type of sampling program that should be used in a production survey of this type.

III. SOLUTIONS

1. *Industrial Hygiene Solution 1.* [hp].

   a. To convert the concentration units to ppmv, first the molecular weight of chlorine must be obtained. Chlorine gas is Cl_2 with 2 gmol of Cl atom per Cl_2 molecule so chlorine gas has a molecular weight of:

   $$Cl_2 \text{ MW} = 2 \ (35.45 \text{ g/gmol}) = 70.9 \text{ g/gmol}$$

   The absolute temperature of the gas is:

   $$T = 10°C + 273.2°C = 283.2 \text{ K}$$

   The pressure of the gas in atmospheres is:

   $$P = 600 \text{ mmHg/760 mmHg} = 0.789 \text{ atm}$$

   Since from the ideal gas law 1 gmol of any gas occupies 22.4 L at 0°C and 1 atm pressure, the concentration of chlorine gas in ppmv can be determined by taking the mass per unit volume value, converting the chlorine gas mass to volume, and then expressing this concentration as volume per million volumes as follows:

   $$Cl_2 \text{ Volume} = \frac{(Cl_2 \text{ Mass}) \ (R) \ (T)}{(\text{MW}) \ (P)}$$

   $$= \frac{(0.015 \text{ g}) \ (0.082 \text{ atm-L/gmol-K}) \ (283.2 \text{ K})}{(70.9 \text{ g/gmol}) \ (0.789 \text{ atm})}$$

   $$= 0.00623 \text{ L} = 6.23 \text{ mL}$$

   Since there are 10^6 mL in a m^3, the concentration in ppmv is expressed as mL/m^3 or $mL/10^6$ mL, or:

   $$\text{Concentration} = (6.23 \text{ mL})/1 \text{ m}^3 = 6.23 \text{ ppmv}$$

   b. Knowing that the concentration of Cl_2 on a volume basis does not change with temperature, at STP there are 6.23 mL = 0.00623 L of Cl_2 in 1 m^3 of air. From the ideal gas law, at STP, these 6.23 mL represent the following mass of Cl_2:

   $$Cl_2 \text{ Mass} = \frac{(Cl_2 \text{ Volume}) \ (P) \ (\text{MW})}{(R) \ (T)}$$

1. *Industrial Hygiene Solution 1.* [hp] (continued)

$$= \frac{(0.00623 \text{ L}) (1 \text{ atm}) (70.9 \text{ g/gmol})}{(0.082 \text{ atm-L/gmol-K}) (273.2 \text{ K})}$$

$$= 0.0197 \text{ g} = 19.7 \text{ mg}$$

The concentration of Cl_2 in mass per unit volume units is then:

Concentration = 19.7 mg/1 m^3 = 19.7 mg/m^3

2. *Industrial Hygiene Solution 2.* [hp]. First, the volume of the smoker's breath that was sampled is calculated from the product of the sampling rate and the duration of sampling as:

Sample volume = (Sampling rate) (Sampling duration)
= (498 mL/min) (159 s) (1 min/60 s)
= 1,320 mL = 0.00132 m^3

The volume of benzene that has been collected is based on the volume concentration of benzene of 10.1 ppbv = 0.0101 ppmv or:

0.0101 ppmv = (Benzene volume, mL)/(0.00132 m^3)
Benzene volume = (0.0101) (0.00132) = 1.33 x 10^{-5} mL
= 1.33 x 10^{-8} L benzene vapor

This volume of benzene represents the following weight based on the ideal gas law:

$$\text{benzene mass} = \frac{\left(\dfrac{624 \text{ mmHg}}{760 \text{ mmHg/atm}} \right) (1.33 \times 10^{-8} \text{ L}) (78 \text{ g/gmol})}{(0.082 \text{ atm-L/gmol-K}) (295.2 \text{ K})}$$

$$= 3.52 \times 10^{-8} \text{ g} = 35.2 \text{ ng}$$

3. *Industrial Hygiene Solution 3.* [pcy]. Permeation is the passage of a chemical through the fabric on a molecular level. Penetration is the passage of a chemical through seams, closures or pinholes. Penetration differs from permeation in that penetration occurs through openings rather than through the material itself.

4. *Industrial Hygiene Solution 4.* [pcy]. EPA recognizes Level A protection for the highest level of skin and respiratory protection; Level B protection for the highest level of respiratory production but a lesser level of skin protection; Level C protection for a lesser level of respiratory protection than Level A or Level B and an equal level of skin protection as Level B; and Level D protection, which provides no protection against chemical hazards.

5. *Industrial Hygiene Solution 5.* [pg]. The correct answer is d. In an industrial environment, the most critical element of any accident prevention program is strong management support. Without this support, the program cannot succeed.

6. *Industrial Hygiene Solution 6.* [pg]. Threshold Limit Values (TLVs) refer to airborne concentrations of substances under which nearly all workers may be repeatedly exposed, day after day, without adverse effects. Control of airborne substances in the work environment is based on the assumption that for each substance there is some safe or tolerable level of exposure below which no significant adverse effects occur. These tolerable levels are called TLVs. The TLVs are published, reviewed, and updated annually by ACGIH. The TLVs are exposure guidelines that have been established for airborne concentrations of many chemical compounds, but are not mandatory federal or state employee exposure standards.

7. *Industrial Hygiene Solution 7.* [pg]. Noise is unwanted sound. The same sound can be perceived as noise depending upon temporal factors such as:

- time of the day
- activities involved
- loudness or intensity of the sound
- duration
- physical condition
- constant or intermittent
- frequency

8. *Industrial Hygiene Solution 8.* [pg]. Sound energy travels as waves through the air and is generated as a result of the oscillation of pressure above and below atmospheric pressure. Excessive or unwanted sound is referred to as "noise". Sound energy interacts with the human system by initially vibrating the tympanic membrane, or ear drum, which conducts the energy to the middle ear where the energy is further conducted, via vibration of the ossicle, into the inner ear. In the inner ear pressure changes in the cochlea and movement of hair cells cause stimulation of the auditory nerve which transmits the signal to the brain. Excessive levels of, and exposure to, sound at a given range of frequencies, combined with a related excess in duration of exposure, can result in damage to the auditory system and concurrent conductive or sensorineural hearing loss. Hearing loss associated only with aging is called presbycusis. Damage to the tympanic membrane or middle ear ossicle can result in conductive hearing loss. Sensorineural hearing loss results when the inner ear or auditory nerve are damaged.

9. *Industrial Hygiene Solution 9.* [pg]. In an industrial environment, exposure to many environmental hazards such as chemical hazards, physical hazards, ergonomic hazards, and biological hazards can produce significant subjective responses or strain on the human body. These environmental hazards are described below.

- Chemical hazards: Inhalation of chemical agents in the form of vapors, gases, fumes, and mists can cause irritation, discomfort, sickness and death in humans. Chemicals may also be absorbed into the body through the skin or can directly damage the skin or eyes.
- Physical hazards: Falling, tripping, exposure to electromagnetic and ionizing radiation, noise, vibration, and extremes of temperature and pressure create physical hazards producing various levels of discomfort and stress in humans.
- Ergonomic hazards: Improperly designed tools or work areas, improper work habits such as improper lifting or reaching, poor visual conditions, and repeated motions in an awkward position can result in ergonomic hazards.
- Biological hazards: Insects, molds, fungi, and bacterial contamination can result in biological hazards in the workplace.

10. *Industrial Hygiene Solution 10.* [pg]. In a fire hazard situation, nothing is absolutely fireproof. Nearly everything can burn given ignition, adequate fuel support and sufficient oxygen. During a fire heat energy is transmitted by convection, conduction, and radiation. Fire, flame, heat, smoke, and toxic gases will spread in a building both vertically and horizontally. More than 75% of fire deaths are caused by smoke and toxic gases. In a fire emergency, very often there are only a few minutes between the beginning of combustion and the development of a dangerous fire.

11. *Industrial Hygiene Solution 11.* [pg]. The four stages of a fire include the following:

1. Incipient stage: During the incipient stage there is no visible smoke, flame nor significant heat. Small combustion particles are being increasingly generated over time. Development of a fire can be either quick or slow; taking either minutes, or hours to days to completely develop.

11. *Industrial Hygiene Solution 11.* [pg] (continued)

    2. Smoldering stage: During this stage the quantity of combustion products increases. Tiny particulate matter increases and becomes visible. Smoke is now visible, but neither flame nor heat are visible.

    3. Flame stage: During this stage flame development begins. Visible smoke decreases and heat production increases. Infrared energy begins to radiate away from the fire and will travel long distances.

    4. Heat stage: During this stage a large amount of heat, flame, smoke, and toxic gases are produced. The transition from the flame to the heat stage is rapid.

12. *Industrial Hygiene Solution 12.* [pg]. Flame, heat, visible smoke, and combustion gases are the product of combustion.

Flames can cause direct burns on contact. Health effects from heat depend on the duration of exposure and the intensity of the fire. Individuals experience increased heart rate, become dehydrated, exhausted, and experience respiratory blockage during exposure to heat from fires. Inhalation of heated air causes blood pressure to drop and capillaries tend to collapse.

Combustion gases are the primary cause of death associated with fires. Health effects due to inhalation of toxic and oxygen deficient gases and smoke depend upon the length of exposure, concentration of the gases, physical condition of the victim, and individual susceptibility. The amount and type of gases present depends upon the chemical composition of the material burning, available oxygen and temperature. Carbon monoxide, hydrogen cyanide, and hydrogen chloride are the three major toxic gases that are produced in a fire situation. Carbon monoxide is the product of incomplete combustion. Hydrogen cyanide is produced by burning natural materials containing nitrogen as well as synthetic material such as silk, acrylonitrile polymers, nylons, polyurethane, and urea-containing resins. Hydrogen cyanide is a rapidly acting toxicant that is 20 times more toxic than carbon monoxide. Hydrogen chloride is formed from the combustion of chlorine compounds, mostly PVC.

13. *Industrial Hygiene Solution 13.* [pg]. Directing and managing an industrial hygiene program is the primary responsibility of an Industrial Hygienist. Examining the work environment, interpreting results of the examination in terms of workplace conditions able to impair health, making specific decisions, and conducting training programs are all part of an industrial hygiene program.

13. *Industrial Hygiene Solution 13.* [pg] (continued)

Industrial hygiene plant survey measures include:

- Preplanning: Preplanning includes obtaining and reviewing process flow sheets, and studying any technical books or publications available on the operation being inspected.
- Recognizing the extent of hazard: This step entails reviewing and listing amounts of raw materials, additives and catalysts, and products used for the entire plant; determining and reviewing the toxicity of the materials of interest and by-products; and making a list of equipment containing design elements likely to be sources of exposures such as leaky pumps, tank vents, product loading stations, etc.
- Preliminary walk-through survey: This step involves using sight and smell senses to identify dust and fumes, gases and vapors existing throughout the plant; noting existing control measures; observing general workroom conditions, movements of workers, approximate time spent in areas of potentially hazardous conditions and employee work habits.
- Survey: This step involves conducting the survey; evaluating the data; making recommendations for improvements; and reporting the findings.

14. *Industrial Hygiene Solution 14.* [pg]. Ergonomics is the study of human characteristics such as capabilities, limitations, motivations, and desires for the appropriate design of the living and work environment. The goals of ergonomic applications are: making the workplace safe, increasing human efficiency, and improving human well-being and the quality of life while at work.

Anthropometry literally means measuring the human. Anthropometry describes the physical dimensions and properties of the body while biomechanics describes characteristics of the human body in mechanical terms. Both are essential aspects of ergonomics used to improve the work environment. Combining anthropometric and biomechanics information has added new dimensions of human factors engineering to classical physiological and behavioral information. This has resulted in improved applications of ergonomic principles to workplace environment problems.

15. *Industrial Hygiene Solution 15.* [pg]. Complaints related to posture (musculoskeletal pain and discomfort) and vision (eye strain, and fatigue) are by far the most frequent health problems

15. *Industrial Hygiene Solution 15.* [pg] (continued)

voiced by computer operators. The postural problems appear to be largely caused by improperly designed and poorly arranged workstation furniture. Successful ergonomic design of the office workstation (furniture, equipment, environment) depends on the proper consideration of several interrelated aspects such as work postures (head and neck, hands and arms, trunk and pelvis, feet and legs) and work activities (visual, motor, auditory, vocal). The design of the workplace and the design of the tasks are interrelated. The work station should be designed to prevent overloading the muscular system, assuring a proper match between the facility and the operator, and aiming for the best mechanical advantage in the design of the task.

The user interfaces with computers through several sensors. The majority of sensory reception occurs through the eyes, as they fix their sight on the monitor, on the source document, and even on the keys. Through individual preferences for the angle of the line of sight, the position of the person's eyes is relatively fixed with respect to the visual target. This eye fixation has rather stringent consequences for neck and trunk postures. The hands are the major output interface between the user and the computer. With current technology, controls and hands are essentially in front of the body and between shoulder and lap height. These have consequences for body posture. Since the controls, such as the keys and mouse, are fixed within the workstation, the operator has no choice but to keep the hands at this location.

A major design concern is to provide the opportunity and means to change body posture frequently during the work period. Maintaining a particular posture, even if it is comfortable in the beginning, becomes stressful as time passes. To permit position changes for the hands/arms and eyes, the input device, i.e., the keyboard, should be movable within the work space. Also, one should be able to adjust the display screen to various heights and angles, which requires an easily adjustable, possible motor or spring driven, suspension system of the support surface. If vision must be focused on the screen as well as on the source document and on the keyboard, all visual targets should be located close to each other, at the same distance from the eyes, and in about the same direction of gaze. Often, the computer screen is arranged too high, forcing the operator to tilt his/her neck severely backward. This causes muscle tension and generates strain on the cervical part of the spinal column. This strain, in turn, regularly leads to complaints about headaches and pains in the neck and shoulder region. Most of these postural complaints can be avoided by proper ergonomic design, proper adjustment, and proper use of the work equipment.

16. *Industrial Hygiene Solution 16.* [pg]. The Occupational Safety and Health Act's (29 CFR 1910.1200) Hazard Communication Standard consists of three elements. They are hazard warning labels, material safety data sheets (MSDSs) and employee training programs. The standard requires chemical manufacturers and importers to make a comprehensive hazard determination for the chemical products they sell. The manufacturers, importers, and distributors must provide information concerning the health and physical hazards of these products. This is accomplished by means of warning labels.

The Hazard Communication Program should include the following elements:

- preparation of a written hazard communication plan
- identification and evaluation of chemical hazards in the work place
- preparation of a hazardous substance inventory
- development of a file of MSDSs
- provision for access to MSDSs for employees
- insurance that incoming products have proper labels
- development of a system of labeling within the facility
- development of a training program
- identifying and training employees who are potentially exposed to hazardous chemicals, and
- evaluation of the programs and redirection where necessary.

17. *Industrial Hygiene Solution 17.* [kg]. Exposure to volatile organic compounds in a workplace may become a health problem based on their concentration levels and exposure times. Often dilution ventilation is required to keep the concentration of volatile organic compounds within acceptable levels.

The dilution air can be estimated using the equation given in the Problem Statement:

$$Q = KV/C_a$$

As defined in the Problem Statement, $K = 5$. The TLV for toluene is 100 ppmv. Since C_a is related to 80% of the TLV, then it is 80 ppmv.

The ideal gas law can be used to calculate the pure contaminant vapor volumetric flow rate as:

$$PV = n\,RT = \frac{M}{MW}\,RT$$

17. *Industrial Hygiene Solution 17.* [kg] (continued)

$$V = \frac{M}{MW}\frac{RT}{P}$$

where M = mass flow rate of contaminant vapor, g/min; MW = molecular weight of the vapor, g/gmol; R = the ideal gas constant, 0.082 atm-L/gmol-K; T = absolute temperature, K; and P = absolute pressure, atm.

The mass flow rate of toluene from the Problem Statement is:

M = (volumetric evaporation rate) (specific gravity)
(unit weight of water)
= (3 gal/8 hr) (0.4 toluene/gal) (0.87) (8.34 lb/gal)
= (1.09 lb/hr) (454 g/lb) (1 hr/60 min) = 8.24 g/min

The resultant toluene vapor volumetric flow rate is:

$$V = \frac{(8.24 \text{ g/min})}{(92 \text{ g/gmol})}\frac{(0.082 \text{ atm-L/gmol-K}) (293 \text{ K})}{(1 \text{ atm})} = 2.15 \text{ L/min}$$

The required diluent volumetric flow rate is then calculated as:

Q = 5 (2.15 L/min)/(80 ppmv) = (10.75 L/min)/(80 L/10^6 L)
= 134,375 L/min (1 ft^3/28.3 L) = 4,748 ft^3/min

18. *Industrial Hygiene Solution 18.* [dks].

a. Using the flow pattern in Figures 13 and 14, the average flow is the total volume divided by the time over which the volume accumulates. The total volume is the area under the curve and is found by slicing the curve horizontally and adding up the area of each slice. Since the flow rates are uniform during the period from 0 to 2 hr and 2 to 4 hr, two slices will do:

$$\overline{Q} = \frac{\text{Total Flow Volume}}{\text{Total Time}} = \frac{\sum_i Q_i \, \Delta t_i}{\sum_i \Delta t_i}$$

$$= \frac{(15 \text{ gpm}) (2 \text{ hr}) (60 \text{ min/hr}) + (7.5 \text{ gpm}) (2 \text{ hr}) (60 \text{ min/hr})}{(2 \text{ hr} + 2 \text{ hr}) (60 \text{ min/hr})}$$

$$= \frac{1,800 \text{ gal} + 900 \text{ gal}}{120 \text{ min} + 120 \text{ min}} = 11.25 \text{ gpm}$$

18. *Industrial Hygiene Solution 18.* [dks] (continued)

    b. Using Figure 14, the average concentration of material X can be estimated in the same fashion as follows:

$$\bar{C} = \frac{\text{Concentration } (\Delta t)}{\text{Total Time}} = \frac{\sum_i C_i \, \Delta t_i}{\sum_i \Delta t_i}$$

$$= \frac{(150 \text{ mg/L}) (2 \text{ hr}) (60 \text{ min/hr}) + (300 \text{ mg/L}) (2 \text{ hr}) (60 \text{ min/hr})}{(2 \text{ hr} + 2 \text{ hr}) (60 \text{ min/hr})}$$

$$= \frac{18{,}000 \text{ mg-min/L} + 36{,}000 \text{ mg-min/L}}{120 \text{ min} + 120 \text{ min}} = 225 \text{ mg/L}$$

    c. The total mass of X from these results is simply the product of the average flow rate and the average concentration, corrected to yield the proper units.

$$M_x = K \, \bar{Q} \, \bar{C}$$

$$= (11.25 \text{ gpm}) (225 \text{ mg/L}) (60 \text{ min/hr}) (4 \text{ hr}) (3.785 \text{ L/gal})$$
$$(1 \text{ kg/10}^6 \text{ mg}) = 2.29 \text{ kg}$$

    d. The total quantity of X being discharged is the sum of the products of the flow rate, concentration, and time interval. The reasoning here is that the mass of compound X in a given unit of volume is simply the concentration times the volume. So, the mass accumulated over the 4-hr time is simply the sum of the individual masses over a uniform time interval of 1-hr each. The calculation is summarized as follows in Table 27.

Table 27
Calculation Table for Question d

i	Q_i (gpm)	C_i (mg/L)	$Q_i C_i$ (gal-mg/L-min)
1	15	150	2,250
2	15	150	2,250
3	7.5	300	2,250
4	7.5	300	2,250
Total			9,000

The total is 9,000 gal-mg/L-min. Converting units the total mass is:

18. *Industrial Hygiene Solution 18.* [dks] (continued)

$$X = (9{,}000 \text{ gal-mg/L-min}) (60 \text{ min/hr}) (3.785 \text{ L/gal}) (1 \text{ kg/10}^6 \text{ mg})$$
$$= 2.04 \text{ kg}$$

    e. The method in Part d uses flow-weighted averages while in Part c simple averages are used. The method used in Part c overestimates the discharge by 12.5% since when the flow is low, the concentration is high and when the flow is high, the concentration is low. If flow and concentration are correlated, simple averages give biased results. Using the result in Part d, the answer to Part b would be given by the total mass divided by the total flow, or:

$$\overline{C} = \frac{\sum_i C_i \, \Delta t_i}{\sum_i \Delta t_i} = \frac{2.05 \text{ kg}}{(2{,}700 \text{ gal}) (3.785 \text{ L/gal}) (1 \text{ kg/10}^6 \text{ mg})}$$
$$= 200 \text{ mg/L}$$

which is less than the 225 mg/L shown in the solution to Part b.

19. *Industrial Hygiene Solution 19* [dks].

    a. Using the flow pattern in Figures 15 and 16, the average flow is the total volume divided by the time over which the volume accumulates. The total volume is the area under the curve and is found by slicing the curve horizontally and adding up the area of each slice. The answer to this question is identical to Part a in Problem 18 above, since the flow rate versus time figures for both problems are identical.

    b. Using Figure 16, the average concentration of material X can be estimated in the same fashion as follows:

$$\overline{C} = \frac{\text{Concentration} (\Delta t)}{\text{Total Time}} = \frac{\sum_i C_i \, \Delta t_i}{\sum_i \Delta t_i}$$

$$= \frac{(300 \text{ mg/L}) (2 \text{ hr}) (60 \text{ min/hr}) + (150 \text{ mg/L}) (2 \text{ hr}) (60 \text{ min/hr})}{(2 \text{ hr} + 2 \text{ hr}) (60 \text{ min/hr})}$$

$$= \frac{36{,}000 \text{ mg-min/L} + 18{,}000 \text{ mg-min/L}}{120 \text{ min} + 120 \text{ min}} = 225 \text{ mg/L}$$

19. *Industrial Hygiene Solution 19.* [dks] (continued)

   c.   The total mass of X from these results is simply the product of the average flow rate and the average concentration, corrected to yield the proper units.

$$M_x = K \, \overline{Q} \, \overline{C}$$

$$= (11.25 \text{ gpm}) \, (225 \text{ mg/L}) \, (60 \text{ min/hr}) \, (4 \text{ hr})$$
$$(3.785 \text{ L/gal}) \, (1 \text{ kg}/10^6 \text{ mg}) = 2.29 \text{ kg}$$

   d.   The total quantity of X being discharged is the sum of the products of the flow rate, concentration, and time interval. The reasoning here is that the mass of compound X in a given unit volume is simply the concentration times the volume. So, the mass accumulated over the 4-hr time is simply the sum of the individual masses over a uniform time interval of 1 hr each. The calculation is summarized as follows in Table 28.

Table 28
Calculation Table for Question d

i	Q_i (gpm)	C_i (mg/L)	$Q_i C_i$ (gal-mg/L-min)
1	15	300	4,500
2	15	300	4,500
3	7.5	150	1,125
4	7.5	150	1,125
Total			11,250

The total is 11,250 (gal-mg/L-min) and converting units the total mass is:

$$X = (11{,}250 \text{ gal-mg/L-min}) \, (60 \text{ min/hr}) \, (3.785 \text{ L/gal}) \, (1 \text{ kg}/10^6 \text{ mg})$$
$$= 2.56 \text{ kg}$$

   e.   Method c now underestimates the discharge by 10% since when the flow is high, the concentration is high and when the flow is low, the concentration is low. Again, if flow and concentration are correlated, simple averages give biased results. Using the result in Part d, the answer to Part b above would be given by the total mass divided by the total flow, or:

19. *Industrial Hygiene Solution 19.* [dks] (continued)

$$\overline{C} = \frac{\sum\limits_{i} C_i \, \Delta t_i}{\sum\limits_{i} \Delta t_i} = \frac{2.56 \text{ kg}}{(2{,}700 \text{ gal}) \, (3.785 \text{ L/gal}) \, (1 \text{ kg}/10^6 \text{ mg})}$$

$$= 250 \text{ mg/L}$$

which is more than the 225 mg/L shown in the solution to Part b.

f. This problem points out that when any correlation (positive or negative) exists between flow and concentration (it usually does!), then a flow-weighted composite sample is needed, unless the variability pattern is known precisely. If flow and concentration are uncorrelated but still variable, composites are also needed. Flow-weighting will not help if samples are randomly correlated, however.

Chapter 10

ISO 14000

Kumar Ganesan

I. INTRODUCTION

Worldwide competition among businesses, coupled with an ever increasing awareness of our environment and costs associated with its maintenance, have led businesses to look at environmental performance as a competitive differentiator. The International Organization for Standardization (ISO) has responded to this need and has formulated an internationally recognized set of environmental management standards. ISO is an internationally recognized, independent standard-setting body. The ISO is composed of member bodies from more than 111 countries. The ISO encourages uniform practices around the world to reduce trade barriers, and is playing a prominent role in encouraging companies to comply with environmental regulations and improve environmental performance.

ISO 14000 is a voluntary standard for Environmental Management Systems (EMSs). It does not require compliance with the regulations of the country in which the company is situated. It is intended, however, to give organizations guidance on the elements of an effective EMS which can be integrated with other management activities and to help organizations achieve their environmental and economic goals. The International Standard specifies the requirements of such an EMS. The success of the system depends on commitment from all levels and functions within an organization, particularly from top management.

Becoming proactive in managing the environmental aspects of an organization will make good business sense. An effective EMS will provide the organization, its directors, its managers, and its employees with several important benefits. The three major areas of benefit are: market advantages, environmental compliance, and cost savings. Based on past experience with ISO 9000 standards designed to improve business products or services, customers will soon consider ISO 14000 certification, designed to improve a company's environmental performance, as a condition of doing business. The organizations certified under ISO 14000 will have a strong differentiating message to deliver to its marketplace, such as: "we are a company proactively improving the environmental performance of our operations and products."

In environmental compliance the implementation of policies, programs, and systems called for under ISO 14000 will allow the organization to effectively manage its operations and products to achieve compliance, and to proactively modify operations and processes to exceed compliance. Cost savings result when an effective EMS is in place because resources are properly allocated, employees are trained and focused, and an EMS stimulates an accurate decision making process, saving time and money. The implementation of life cycle analysis under ISO 14000 provides a tool that identifies environmental issues throughout a product's life. Eliminating or reducing these issues means reducing or eliminating the activities associated with them, which translates into cost savings.

Under ISO 14001 the basic requirements of an EMS consist of environmental policy planning, implementation, operation, checking, corrective action, and management review. These are the ISO standards by which the organizations/facilities should establish their EMS and the major areas audited for certification.

An environmental policy is the backbone of the system. Top management should define the organization's environmental policy and should ensure that: it is appropriate to the nature, scale, and environmental impacts of activities, products and services provided by the organization. The environmental policy provides a framework for setting and reviewing environmental objectives and targets; includes a commitment to continual improvement, prevention of pollution, and compliance with relevant environmental legislation and regulations. The policy should be documented, implemented, maintained and communicated to all employees and should be available to the public.

An organization can be certified through third party auditing. This procedure involves identifying an accredited registrar firm for certification. The certified auditors from the registrar's office then will plan to visit the facility and audit the EMS of the company requesting the audit. The audit team then completes its audit report and submits its findings to the registrar. The registrar, based on the findings and recommendations of the audit report, makes a registration decision and acts accordingly. The registrar communicates the decision to the organization that they audited. Effective communication between the registrar and the facility to be audited well before the time of audit will alleviate many problems and make the audit go smoothly and successfully. A pre-audit by the facility is essential to identify problem areas and fix them before the third party audit begins. An organization also can self-declare, without the third party audit.

This chapter in this workbook is intended to provide an overview of ISO 14000. There are 16 questions and answers spanning topics from simple definitions to complex auditing procedures. The reader is advised to refer to books published on the subject to further expand the understanding of ISO 14000. Two publications recommended for reference purposes are listed in the reference section below.

REFERENCES

1. **Tibor, T. and Feldman, I.**, *A Guide to the New Environmental Management Standards*, Irwin Professional Publishing, Chicago, Illinois, 1995.
2. **von Zharen, W. W.**, *Understanding the Environmental Standards*, Government Institutes, Inc., Rockville, Maryland, 1995.

II. PROBLEMS

1. *ISO 14000 Problem 1*. (ISO, ANSI). [pcy]. What is the ISO?

2. *ISO 14000 Problem 2*. (ISO, goals). [hb]. Briefly discuss the purpose and goals of the ISO.

3. *ISO 14000 Problem 3*. (ISO, ISO 14000). [hb]. What is ISO 14000?

4. *ISO 14000 Problem 4*. (ISO, ISO 14000, compliance). [hb]. Discuss the relationship of ISO 14000 to regulatory compliance.

5. *ISO 14000 Problem 5*. (ISO 14000, global market standards). [car]. You have been hired as the environmental officer for a manufacturing company that intends to enter the global market. As part of the business plan for the company entering the global market, the company intends to become ISO 14000 certified.
    What are the major components of ISO 14001 EMS that must be implemented by the company in order to pass the audit requirement for certification?

6. *ISO 14000 Problem 6*. (ISO 14000). [sn]. List and discuss the benefits and pitfalls of the ISO 14000 series of standards to industry from an international point of view.

7. *ISO 14000 Problem 7*. (ISO 14000). [sn]. Read the following statements regarding the ISO 14000 series of standards. Carefully justify your answers as TRUE (T) or FALSE (F).

    a. ISO 14000 standards are based on a principle assumption that better environmental management will lead to better environmental performance, increased efficiency, and a greater return on investment. The standards do not explicitly indicate how to achieve these goals, nor prescribe what environmental performance standards an industry must achieve.
    b. ISO 14000 standards are regulatory standards developed by the International Organization for Standardization (ISO).
    c. ISO 14000 standards are market-driven and therefore are based on voluntary involvement of all interests in the marketplace.
    d. ISO 9000 and 14000 standards share the following generic management elements: 1) setting policies; 2) establishing document

7. *ISO 14000 Problem 7.* [sn] (continued)

   control, training, corrective action, management review, and continual improvement; and 3) controlling operational processes. The main difference is that ISO 9000 is designated for system quality while ISO 14000 is targeted at environmental management.

 e. The adoption of ISO 14000 is a one-time commitment. The company, however, needs to renew the certificate yearly.

 f. Currently there are five certification bodies in the U.S. for ISO 14000 standards including the American National Standards Institute (ANSI), the Registrar Accreditation Board (RAB), American Society for Testing and Materials (ASTM), the American Society for Quality Control (ASQC), and the Environmental Auditing Roundtable (EAR).

 g. A main driving force of ISO 14000 standards is the need of U.S. EPA for the replacement of an obsolete regulatory system.

 h. A minimum education requirement for an auditor with 5 years "appropriate work experience" is a high school diploma or equivalent.

 i. Companies can only demonstrate compliance through third-party registration.

 j. A single ISO certificate can cover several sites or facilities, or portions of sites or facilities within a single company.

8. *ISO 14000 Problem 8.* (ISO, environmental management). [kg]. What is an environmental management system?

9. *ISO 14000 Problem 9.* (audit, certification). [kg]. What are the major areas to be audited for certification under ISO 14001?

10. *ISO 14000 Problem 10.* (certification, registration). [kg]. Explain the process involved in the certification (registration) of a company under ISO 14001.

11. *ISO 14000 Problem 11.* (audit plan, auditors). [kg]. What are the major items that must be included in an audit plan?

12. *ISO 14000 Problem 12.* (Lead Auditor, auditee). [kg]. Explain the purpose of the opening and closing meeting with a third-party audit team.

13. *ISO 14000 Problem 13.* (objective evidence, interviewing). [kg]. The responsibility of the audit team is to gather objective evidence. Explain the methods and approaches the auditors may use to gather this evidence.

14. *ISO 14000 Problem 14.* (conformance, non-conformance, major non-conformance, minor non-conformance, observation). [kg]. Explain the term non-conformance and the reasons for non-conformance under ISO 14001.

15. *ISO 14000 Problem 15.* (policy, continual improvement, pollution prevention). [kg]. The policy statement of a facility is an important part of the ISO 14000 process. What are the major elements that must be included in a facility's environmental policy statement?

16. *ISO 14000 Problem 16.* (audit, environmental law, non-conformance). [kg]. A third-party audit is conducted in a pulp and paper mill facility. You are one of the auditors assigned to verify written procedures that are in place and thus to verify the effectiveness of the facility's environmental management system. One of the personnel you are interviewing is the facility's environmental manager. You asked the manager to verify that she has access to all of the regulations and laws that are applicable to the plant. The manager signals you to follow her to a library in the next room. She stops in the library and proudly points toward the four bookcases occupied with texts on environmental laws.

You are impressed but you continue to ask the environmental manager another question. "Can you show me the procedure you use to evaluate regulatory compliance within your facility?" The manager immediately responds by saying that she does not need a procedure for compliance evaluation because she is intimately familiar with what documentation needs to be reviewed when a compliance issue arises.

Discuss the above situation with regard to conformance and non-conformance.

III. SOLUTIONS

1. *ISO 14000 Solution 1.* [pcy]. ISO is the International Organization for Standardization, a worldwide federation founded in 1947 to promote the development of international manufacturing, trade, and communication standards. ISO is composed of member bodies from more than 111 countries. The American National Standards Institute (ANSI) is the U.S. representative to ISO.

ISO receives input from government, industry, and other interested parties before developing a standard. All standards developed by ISO are voluntary; no legal requirements force countries to adopt them. However, countries and industries often adopt ISO standards as requirements for doing business, thereby making them virtually mandatory in these cases.

ISO develops standards in all industries except those related to electrical and electronic engineering. Standards in these areas are developed by the Geneva-based International Electrotechnical Commission (IEC), which has more than 40 member countries, including the U.S.

(Source: **CEEM Information Services**, *A Guide to ISO 14000*, ASQC Quality Press, Milwaukee,Wisconsin, 1996.)

2. *ISO 14000 Solution 2.* [hb]. The purpose and goals of the International Organization for Standardization is to improve the climate for international trade by leveling the playing field. The concept is that by encouraging uniform practices around the world, barriers to trade are reduced. If the management processes of companies in any country could be compared more readily with the management processes of companies in any other country then international trade would be made simpler and less risky. Decisions as to which product to purchase could be more readily made based on the competitive edge that a particular product or company possessed rather than on the peculiarities of the way things are done in that country.

3. *ISO 14000 Solution 3.* [hb]. ISO 14000 is the International Organization for Standardization's standard for quality Environmental Management Systems. ISO 14000 describes in considerable detail what a company must do without prescribing how it must be done. When complete, ISO 14000 will be comprised of approximately 20 components. It will be sufficiently specific so that it will be possible to audit companies for their conformance with the standard. Certification of conformance is also available.

Examples of the components of the ISO 14000 Environmental Management Systems are:

3. *ISO 14000 Solution 3*. [hb] (continued)

- Environmental Management Principles
- Environmental Labeling
- Environmental Performance Evaluation
- Life Cycle Assessment
- Principles of Environmental Auditing
- Terms and Definitions

4. *ISO 14000 Solution 4*. [hb]. ISO 14000 is a voluntary standard for Environmental Management Systems. It does not require compliance with the regulations of the country in which the company is located. In some countries, the regulations may be more stringent than the standard. It seems likely, however, that in some countries achieving certification of adherence to the standard would improve the quality of environmental practices in that country. If, as expected, many countries adopt laws that require imported products to have been produced by companies certified to be adhering to ISO 14000 then environmental practices will almost certainly be improved worldwide. The impact on the level of compliance with environmental regulations is, however, as yet unclear.

5. *ISO 14000 Solution 5*. [car]. The ISO 14000 series of standards is a mechanism for managing environmental impacts by companies that do business around the world. The ISO 14000 series of standards is divided into two categories:

1. Organizational evaluation
   - ISO 14001 - environmental management systems (EMS)
   - ISO 14010/14011/14012 - environmental auditing
   - ISO 14031 - environmental performance evaluation (EPE)
2. Product evaluation
   - ISO 14040/14041/14042/14043 - life cycle assessment (LCA)
   - ISO 14020/14021/14024 - environmental labeling
   - ISO 14060 - environmental aspects in products

The principal elements of an ISO 14001 EMS that must be implemented by the company should include the following:

5. *ISO 14000 Solution 5.* [car] (continued)

- Commitment and Policy.
- Planning.
- Implementation.
- Measurement, Monitoring and Evaluation.
- Review and Improvement.

6. *ISO 14000 Solution 6.* [sn]. There are a number of possible benefits and pitfalls of the ISO 14000 standards. The following are some examples.

Benefits:
a. The ISO 14000 standards provide industry with a structure for managing their environmental problems which presumably will lead to better environmental performance.
b. It facilitates trade and minimizes trade barriers by harmonization of different national standards. As a consequence, multiple inspections, certifications, and other conflicting requirements could be reduced.
c. It expands possible market opportunities.
d. In developing countries, ISO 14000 can be used as a way to enhance regulatory systems that are either nonexistent or weak in their environmental performance requirements.
e. A number of potential cost savings can be expected, including:
   - increased overall operating efficiency
   - minimized liability claims and risk
   - improved compliance record (avoided fines and penalties)
   - lower insurance rates.
f. ISO 14000 registration can demonstrate an organization's commitment and credibility regarding environmental issues (e.g., corporate image and community goodwill).

Pitfalls:
a. Implementation of ISO 14000 standards can be a tedious and expensive process.
b. ISO 14000 standards can indirectly create a technical trade barrier to both small businesses and developing countries due to limited knowledge and resources (e.g., complexity of the process and high cost of implementation, lack of registration and accreditation infrastructure, etc.).

6. *ISO 14000 Solution 6.* [sn] (continued)

    c.   ISO 14000 standards are voluntary. However, some countries may make ISO 14000 standards a regulatory requirement which can potentially lead to a trade barrier for foreign countries who cannot comply with the standards.

    d.   Certification/registration issues, including
- the role of self-declaration versus third party auditing
- accreditation of the registrars
- competence of ISO 14000 auditors
- harmonization and worldwide recognition of ISO 14000 registration.

7. *ISO 14000 Solution 7.* [sn].

    a.   True. <u>Explanation</u>: ISO 14000 standards are process standards, not performance standards. They do not prescribe to a company what environmental performance they must achieve. They provide a building block for a system to achieve environmental goals. As a consequence, these standards will lead a company to cost savings through better performance of the environmental aspects of an organization's operations.

    b.   False. <u>Explanation</u>: ISO 14000 standards are international, voluntary standards developed by Technical Committee 207 (TC 207) of the International Organization for Standardization (ISO).

    c.   True. <u>Explanation</u>: ISO 14000 standards are market-driven and therefore are based on voluntary involvement of all interests in the marketplace.

    d.   True. <u>Explanation</u>: ISO 9000 quality standards were developed to address quality management. The ISO 14000 standards are emerging to address a similar need in the environmental area. Both standards provide basic elements of an effective management system for a company to achieve specific goals.

    e.   False. <u>Explanation</u>: The adoption of ISO 14000 standards is a continual commitment. Top management must establish the company's environmental policy and make a commitment to continual improvement and prevention of

7. *ISO 14000 Solution 7.* [sn] (continued)

pollution and to comply with relevant environmental legislation and regulations. Once the company is certified, the certificate is normally valid for 3 years. This may vary depending upon the certification body. The certification body must conduct surveillance audits no less frequently than once a year and carry out a full audit after 3 years.

f. False. <u>Explanation</u>: As of July 1996, there were only two certification bodies in the U.S. for ISO 14000 standards including the American National Standards Institute (ANSI) and the Registrar Accreditation Board (RAB).

g. False. <u>Explanation</u>: U.S. EPA has been participating actively in the standards development process. At present there is no indication of adoption of the ISO 14000 standards as a possible regulatory requirement. The main driving force for the ISO 14000 series of standards is mainly from the private sector.

h. True. <u>Explanation</u>: General qualification criteria for environmental management system auditors include education and work experience. Auditors should have completed at least a secondary education or equivalent with 5 years "...appropriate working experience or a college degree with 2 years appropriate work experience".

i. False. <u>Explanation</u>: Companies can demonstrate compliance through either a self-declaration or third-party registration.

j. True. <u>Explanation</u>: Under ISO 14001 certification, a single certificate can cover a specific site of a company, a specific facility, several facilities or portions of sites or facilities. For example, one ISO 14001 certificate might encompass four different sites of a company in four different states and a portion of a site in a fifth state if these sites are audited at the same time against the same standard.

8. *ISO 14000 Solution 8.* [kg]. The environmental management system (EMS) of ISO 14001 is that part of the general management system that includes organizational structure, planning activities, responsibilities, practices, procedures, processes, and resources for developing, implementing, achieving,

8. *ISO 14000 Solution 8.* [kg] (continued)

reviewing and maintaining the environmental policy of an organization. It is a structured process for the achievement of continual improvement related to environmental matters. The facility has the flexibility to define its boundaries and may choose to carry out this standard with respect to the entire organization, or to focus the EMS on specific operating units or activities of the organization.

The EMS enables an organization to identify the significant environmental impacts that may have arisen or that may arise from the organization's past, existing or planned activities, products or services. It helps the organization to identify relevant environmental, legislative, and regulatory requirements that may be imposed on it. Finally, the EMS helps in planning, monitoring, auditing, corrective action, and review activities to assure compliance with established policy and allows a company to be proactive in terms of meeting anticipated new standards and compliance objectives.

9. *ISO 14000 Solution 9.* [kg]. The five major areas audited for certification include the following:

    a. Environmental policy.
    b. Environmental planning.
    c. Implementation and operation.
    d. Checking and corrective action.
    e. Management review.

These are the ISO standards by which the facilities shall establish their environmental management system (EMS).

The environmental policy is the backbone of the EMS. The policy must reflect the commitment of top management and becomes the basis upon which the organization sets its objectives and targets. The policy statement should be clear and be capable of being understood by internal and external interested parties. The policy should be periodically reviewed and revised as necessary to reflect changing conditions within and outside the organization.

Environmental planning is the second major audited area for certification. Environmental planning has four elements:

    • Environmental aspects
    • Legal and other requirements
    • Objectives and targets
    • Environmental management program

The organization must identify significant environmental aspects that should be addressed as its priorities in the EMS. The

9. *ISO 14000 Solution 9*. [kg] (continued)

process considers, for example, the emission of contaminants to multimedia or the use of raw material as a measure used to identify significant environmental aspects of the operation of a business.

Implementation and operation consist of seven elements:

- Structure and responsibility.
- Training, awareness and competence.
- Communications.
- Environmental documentation.
- Document control.
- Operational control.
- Emergency preparedness and response.

Thus, the organization should establish a system by which environmental responsibilities are not just confined to the environmental function of the company, but are included in other areas of the organization, such as operations.

Checking and corrective action includes four elements:

- Monitoring and measurement.
- Non-conformance and corrective and preventive action.
- Records.
- Environmental system audit.

The organization shall develop procedures to identify, maintain and update records related to EMS. Internal audits or third-party audits should be conducted impartially and objectively.

The management review is designed to monitor continual improvement, suitability and effectiveness of the EMS. The scope of the management review should be comprehensive and the observations, conclusions and recommendations developed from the review should be documented for necessary action.

10. *ISO 14000 Solution 10*. [kg]. There are some preliminary steps considered before the registration process begins. The organization must have implemented and have in operation for 3 to 6 months an environmental management system (EMS) conforming to the requirements of ISO 14001. The organization should have the EMS documentation required to verify that it conforms to the requirements of ISO 14001. The personnel whose job activities involve significant environmental impacts must have received adequate training. The internal audit system must be fully operational and proven to be effective. The EMS must have gone through at least one management review.

10. *ISO 14000 Solution 10.* [kg] (continued)

With these preliminary items in order the organization then can identify/select an accredited registrar firm for certification. Once a registrar is selected, the organization will be asked to complete an application that includes questions about the size of the organization, the scope of its operations and the status of its existing EMS. The application will also explain the rights and responsibilities of both the registrar and the auditee (the organization that is requesting certification). The application also indicates the auditee's right to confidentiality, to voice complaints and to appeal the outcome of the registration audit.

A pre-audit is optional in the U.S. However, conducting a pre-audit may help to decide the organization's state of readiness for the registration audit. A full, on-site registration audit is conducted by the registrar's audit team for a thorough examination of the organization's EMS. The findings of the audit team will be revealed during a formal closing meeting with senior management representatives of the organization. The lead auditor prepares the audit report with the help of the audit team and delivers it to the registrar. The registrar, based on the recommendations and findings in the audit report, makes a registration decision, acts accordingly, and communicates the decision to the organization that was audited.

11. *ISO 14000 Solution 11.* [kg]. Auditing a facility by a third-party for its certification involves several steps. Proper planning and management are very essential for effective auditing. The lead auditor must prepare an audit plan to ensure a smooth audit process. The audit plan must, in general, remain flexible so that any changes to the audit that are found necessary during the actual audit process can be made without compromising the audit.

An audit plan must include the following items:

- A stated scope and objectives for the audit. This includes the reason for conducting the audit, the information required and the expectation of the audit.
- Specification of the place, the facility, the date of the audit, and the number of days required to perform the audit.
- Identification of high priority items of the facility's and/or organization's EMS.
- Identification of key personnel who will be involved in the auditing process.
- Identification of standards and procedures (ISO 14001) that will be used to determine the conformance of various EMS elements.

11. *ISO 14000 Solution 11.* [kg] (continued)

- Identification of audit team members including their special skills, experience and audit background.
- Specification of opening and closing meeting times.
- Specification of confidentiality requirements during the audit process, i.e., who will have access to the audit results, requirement that employees questioned must feel free to give honest answers to ensure a comprehensive and impartial audit.
- Specification of the format of the audit report, the language, distribution requirement and the expected date of issue of the final report.
- Identification of safety and related issues associated with entry and inspection of various portions of the facility, along with other equipment required to conduct an effective and efficient audit

An audit plan that is complete and well thought out will ensure a smooth audit process. Thus, the audit plan plays a major role in the certification process.

12. *ISO 14000 Solution 12.* [kg]. In a formal third-party audit (after the audit team arrives at the facility), the first order of business is to conduct an opening meeting. After the completion of the audit, the team will conduct an exit or closing meeting. These meetings are formal, especially during third-party audits for certification. The lead auditor is responsible for conducting the meeting.

A brief description of the contents of an opening and a closing meeting is given below.

The opening meeting is chaired by the lead auditor (LA). The LA introduces the audit team to facility management. The scope, objective, and the audit plan are reviewed and all parties agree on the proposed timeline contained in the audit plan. An audit time table, including shifts, is discussed for the execution of the audit. The resources necessary to successfully complete the audit are discussed and confirmed by the auditee. A clear line of communication is established between the audit team and the facility. The areas and locations within the facility and the approximate time the auditors will be examining each are established. Confidentiality requirements, site safety and emergency procedures, guides to accompany the auditors, and the date and time of the closing meeting are additional items that are

12. *ISO 14000 Solution 12.* [kg] (continued)

discussed in the opening meeting. The opening meeting also provides the opportunity for management to clear up any confusion or questions they have about the audit. The attendance and minutes are kept as part of the record of the facility audit.

The closing meeting is conducted after the completion of the audit and before the audit team leaves the facility. The main purpose of this meeting is to present audit observations to senior management of the facility. Again, the LA conducts the closing meeting. It is important to compliment facility personnel for their cooperation and their understanding of the importance of the audit visit. The LA states the disclaimer that the audit is a snapshot of events in time and may have missed some items that are in non-conformance, or may have missed evaluating some of the existing procedures. With this disclaimer the LA invites each audit team member to summarize their findings and describe any non-conformance or other observations they have related to the audit. The LA then presents an overall summary, followed by the team's conclusions and recommendations. Facility management has the opportunity to ask questions and to clarify the findings and to agree on dates for corrective actions as necessary. The LA discusses any follow-up activities and closes the meeting. The attendance and minutes are kept as part of the record of the facility audit.

The LA provides a copy of the preliminary report with the findings to the facility's management and obtains their signature for the receipt of the report.

13. *ISO 14000 Solution 13.* [kg]. During a third-party audit, the auditors are sent to their assigned work stations to observe and evaluate the company's environmental management system (EMS). The main responsibility of the audit team is to collect objective evidence, which is any information that can be verified, regarding the conformance of this EMS to ISO 14001 standards. Objective evidence can be gathered using a number of methods such as examination of EMS documents, inspection of documents, products, procedures and interviews with individuals. The communication skills of the auditor are very important in order to obtain meaningful results from interviews with individuals. It is the auditor's job to make sure that the individual being interviewed cares enough about the message to understand the message, confirm it, and act upon it. When interviewing, questions asking how, why, when, where, what, and who seem to work well in soliciting meaningful answers from the facility staff. Asking crunch questions like show me how, show me what you do, etc., are also effective in gathering valuable information.

Thus, objective evidence includes verifiable information, and records or statements of facts. It can also be based on interviews,

13. *ISO 14000 Solution 13.* [kg] (continued)

examination of documents, observation of activities and conditions, results of measurements, tests conducted within the scope of the audit, etc. The auditors can use qualitative or quantitative evidence to determine the effectiveness of the EMS.

14. *ISO 14000 Solution 14.* [kg]. A non-conformity is a deficiency present in an environmental management system (EMS) that is being audited. Non-conformities are classified as either major or minor. The auditor has to gather objective evidence, with necessary documentation that is verifiable, to document that an EMS fulfills the requirements set out by ISO 14001. The major non-conformity can be due to insufficient EMS procedures or improper implementation of the EMS or a single deficiency in the EMS, an activity, product or service. In addition, a series of minor non-conformities that indicate the failure of an effective EMS will result in a major non-conformity.

Non-conformity does not mean the facility cannot be certified. The facility will be given time and opportunity to correct the non-conformity. Once the facility corrects the problems identified in the non-conforming areas then they can contact the registrar and show evidence of compliance. Based on a review of this new information the registrar can recommend certification for the facility.

Non-conformities can be classified as minor if the facility has an overall acceptable level of existing EMS but lapses in following all of the EMS requirements and documentation. In general, minor non-conformities are viewed as activities that do not directly affect the environmental performance of products or services and could be corrected easily.

Several minor non-conformities may be concluded to represent a single major non-conformity if the minor non-conformities are indicative of poor or inadequate implementation of the EMS at the facility being audited.

15. *ISO 14000 Solution 15.* [kg]. A company's environmental policy represents the organization's mission, vision, and commitment to environmental management. The policy provides a statement of the general direction the company is going to take regarding environmental management. It is the responsibility of top level management to create an environmental policy that meets ISO 14001 standards and requirements. Besides the requirements set forth in ISO 14001, the ISO document 14004 also gives general guidelines on principles, systems and supporting techniques related to a company's environmental policy.

The environmental policy of an organization must include several elements such as:

15. *ISO 14000 Solution 15.* [kg] (continued)

- The setting of environmental objectives and targets for the organization.
- The advocacy of the use of best technology and management practices.
- The commitment to meet or exceed any legal and regulatory requirements related to environmental compliance for the company.
- The commitment to continual improvement and prevention of pollution.
- The framework for setting and reviewing environmental objectives and targets of the company.
- The commitment to consider the views of interested parties.

The policy must be documented, implemented and maintained, and must be communicated to all employees. The policy should also be available to the public.

16. *ISO 14000 Solution 16.* [kg]. This is a major non-conformance because the facility lacks a procedure to evaluate its compliance with applicable regulations. The ISO 14001 standard, Section 4.2.2, states that "The organization shall establish and maintain a procedure to identify and have access to legal and other requirements to which the organization subscribes directly applicable to the environmental aspects of its activities, products or services." Thus, the facility is in non-conformance with Section 4.2 and Clause 4.2.2.

The issuance of non-conformance does not mean that the facility cannot be certified. Since this is a major non-conformance, however, the facility must take corrective action and show that the procedure is in place and effective in order to come into conformance with ISO 14001 standards. Having all this in place, the facility may request that the registrar reconsider its certification.

Chapter 11

NATIVE AMERICAN ISSUES

Terry Baxter and Robert Tidwell

I. INTRODUCTION

Native American Indian reservations represent approximately 4% of all of the lands within the U.S. Some of these reservations, originally thought to be land of little value, actually contain large uranium, coal, oil or timber resources having significant economic value. Through the Bureau of Indian Affairs (BIA), the federal government has made leases or contracts available to non-Indians for ranching and for extraction of raw materials from reservation lands. While the development of these resources has provided some short-term economic benefit to the tribes, the environmental and cultural damages from the extraction, exploitation, and mismanagement of these resources have been significant. Many reservation lands experience and have to address the impacts of water pollution, water diversion, habitat degradation, and/or extensive soil erosion from overgrazing due to this development. More recently, impacts that accompany the development of gaming facilities and increased visitors are also occurring.

Historically, two primary factors have caused the development of resources on and off the reservation to be so damaging to the environment and a tribe's culture. First, tribes have had little if any jurisdictional authority over their lands regarding environmental protection, and they have had little legal oversight of potentially environmentally and culturally damaging activities off the reservation. It has only been since the 1970s that tribes themselves could make their own decisions regarding economic development and the management of resulting environmental and social impacts. Second, the technical training and expertise necessary for understanding the potential long-range implications and impacts of economic and resource development on the reservation's environment and culture has also been lacking within the tribes. Although both of these deficiencies are currently in the process of being reversed, the environmental issues Native Americans must face are complex and tied to traditional cultures and spiritual beliefs as well as to concerns of a depressed tribal economy.

In 1984 the U.S. EPA implemented the Policy for the Administration of Environmental Programs on Indian Reservations which sought to enable tribal involvement in environmental regulations and EPA's programs. Two important features of this policy established that either the U.S. EPA or the tribes should have primacy regarding environmental laws on reservations or Indian lands, not the states. In addition, under the authorization of Congress, the EPA assists tribes in developing and implementing tribal environmental programs. Beginning in 1986, amendments to several major environmental laws clearly enumerated the conditions under which EPA can delegate program authority to tribes, thus helping to clarify some of the previously uncertain and contentious debates regarding jurisdictional right between states and tribes. Prior to such amendments, the debates over whether states or tribes had jurisdiction on reservations for the enforcement of environmental regulations often left reservations in a "regulatory hole," and vulnerable to companies looking to exploit this lack of clear authority over environmental regulations. Currently, EPA can delegate specific enforcement and regulatory authority to Indian tribes, as is similarly done for states in many environmental programs. Once obtained, this delegated authority positions tribes to be treated as sovereign entities, as are states, with regard to environmental laws.

Generally any plans for economic development are often evaluated for possible infringement on the tribe's sovereignty and compatibility with traditional cultural values. However, tribal economies in serious need of economic development and employment opportunities may find it difficult to resist the allure of revenues that would be generated by allowing waste disposal sites to be located and operated within reservation boundaries. Hazardous waste incinerators have been accepted by tribes (Kaibab-Paiutes, Kaw), as have landfills (Campo Band, Rosebud Sioux). Still other tribes have considered allowing nuclear waste storage facilities to be operated on their reservations. Although these decisions do address the concerns of economic development, many fear that the long-term potential for environmental and cultural damage will not be worth short-term economic gains.

Tribes must now decide what level of jurisdiction they are either willing or able to assume with regard to the environmental regulation of activities on the reservation. In addition, tribes must become educated and trained in many areas of expertise to better participate in the process of solving existing environmental problems as well as managing emerging ones so that they are able to protect the health and culture of the tribe, while at the same time stimulating needed economic development on the reservations.

The problems contained in this chapter are not presented with the intent that the perspective of the Native American will be fully understood. Rather, it is hoped that these problems will provide non-Indians with a greater awareness of the difficult and complex nature these issues represent. In addition, it is hoped that Native Americans will find value in these problems and use them as a catalyst for reexamining their identity with Mother Earth. Most of these problems are well suited for conducting in-class discussions and role-play exercises.

II. PROBLEMS

1. *Native American Issues Problem 1*. (microbial degradation, hydrocarbons, underground storage tanks). [lrh]. In the 1990s Native American tribes began working with EPA to monitor underground storage tanks (USTs) to comply with RCRA amendments enacted in 1984. As a result, a leaking tank was detected at a gasoline service station located on a tribe's reservation. Since product from the tank reached the groundwater table, this leak became a major health risk issue. The cost incurred by the tribe just to begin the clean-up process was over $1,000,000.

    a.   Through a library or Internet search find out what Reservation this leak occurred on.

    b.   Once it had been established that the UST was leaking, what are some of the actions that the tribe could implement to determine the extent of contamination?

    c.   Assuming that the soil around the UST is contaminated with gasoline hydrocarbons (benzene, toluene, ethylbenzene and xylenes), explain how the tribe might use microbial degradation to clean up the contaminated soil.

2. *Native American Issues Problem 2*. (microbial degradation, Native American culture). [lrh] Select one Native American tribe i.e., Lakota, Iowa, Dakota, etc., and study its traditional culture. Discuss how the use of microbial degradation either might or might not conflict with that tribe's cultural way of life.

3. *Native American Issues Problem 3*. (seventh generation, circle of life). [lrh]. According to Caduto and Bruchac (*Keepers of Life, Discovering Plants Through Native American Stories, and Earth Activities for Children*, Fulcrum Publishing, Golden, Colorado, 1994) many Native American cultures believe that all things (plants, animals, rocks, air, water, soil, humans, etc.) are related and depend on one another for their existence. The Earth is a living being; it is our Mother. Humans need to live within and honor the circle of life by giving and receiving. When they take from Mother Earth, they must give back to her.

Many Native American tribes believe that each generation has influence over the next seven generations. We need to think deeply about how our actions today will affect the next seven generations. The seventh generation of Native Americans to live on Federal reservations are being born into a world of environmental problems.

    a.   Can today's Native American reservation resident make a cleaner environment for the seventh generation after them?

3. *Native American Issues Problem 3.* [lrh] (continued)

    b. What skills will they need to prepare for this task?

Reference: **Caduto, M. J. and Bruchac, J.**, *Keepers of Life, Discovering Plants Through Native American Stories, and Earth Activities for Children*, Fulcrum Publishing, Golden, Colorado, 1994.

4. *Native American Issues Problem 4.* (jurisdiction). [lrh]. Using the Internet determine if Native American tribes are under the jurisdiction of the U.S. EPA or the state they are located in to ensure compliance with the Clean Water Act. Can Native American tribes develop their own codes and regulations to manage waste disposal and treatment to ensure clean water for their people under U.S. EPA guidelines?

5. *Native American Issues Problem 5.* (Native American, U.S. EPA, regulations). [lrh]. Using a specific regulatory program, explain how Native American tribes are handled by U.S. EPA regulations?

6. *Native American Issues Problem 6.* (Native American, U.S. EPA, regulations, states). [lrh]. How are Native American tribes handled differently by state regulations versus U.S. EPA regulations?

7. *Native American Issues Problem 7.* (Native American, Pine Ridge Reservation, insectary, beneficial, noxious weeds). [lrh]. A Native American from the Pine Ridge Reservation has established an insectary to market beneficial insects for the control of noxious weeds. The local reservation land committee has authorized the use of insecticide in the area near the insectary to control a grasshopper out-break. Local ranchers support the use of insecticides for this purpose. The tribal land committee has asked the Bureau of Indian Affairs (BIA) and USDA's Animal and Plant Health Inspection Service (APHIS) to help handle the application of the insecticide. If the insecticide drifts onto the insectary it can be put out of business. Assume that you are the owner of the insectary.

    a How can the insectary be protected?

    b. If damage to the insect population does occur, what type of legal or non-legal action can be taken to recover damages?

8. *Native American Issues Problem 8.* (burial site, landfill). [lrh]. An ancestral Native American burial site has been found on the location the tribe has selected for a landfill. It has taken 5 years to obtain approval for this landfill site through the required tribal government procedures and public referenda. Approval was obtained prior to discovering the burial site.

    a. Comment on whether the tribe should continue to develop this site and ignore their ancestors, or begin the process of selecting a new site and face Federal government fines for not complying with the Resource Conservation and Recovery Act (RCRA).

    b. Tribal leaders have selected you to explain this issue to the general reservation population. How would you represent this conflict to the general tribal population?

9. *Native American Issues Problem 9.* (burial site, landfill). [teb]. An environmental consultant unfamiliar with a tribe's cultural and religious traditions has been hired to assist the tribe in finding a solution to the landfill sitting problem described in the previous problem. Using a Euro-American perspective, what would this consultant likely recommend in regard to the burial ground? How might this recommendation create controversy and conflict with ancestral tribal cultural values?

10. *Native American Issues Problem 10.* (Native American, economic development, cultural heritage, mining). [lrh]. You have been elected as Tribal Chairperson for a Native American tribe in southwestern South Dakota. Many of the tribe's members are on welfare. Due to changes in Federal regulations the length of time that welfare recipients can receive payments will soon change and there are few employment opportunities that currently exist on the Reservation. A mining company has approached you requesting permission to mine a clay mineral called zeolite used in water and wastewater treatment applications. This company promises to hire local Native Americans, which could help ease the effect of the loss of welfare benefits for some tribal members. The location of the zeolite is in a very arid part of the reservation which has been used mainly for sacred ceremonies, cattle grazing, geological study, and recreation. As the new Tribal Chairperson, you need to make a decision as to what is best for all tribal members. What type of information do you need from the tribe's natural resource staff to help you make this decision? Be specific to this region of South Dakota.

11. *Native American Issues Problem 11*. (Native American, public safety, food contamination, risk, risk communication). [lrh]. The manager of a retail grocery/hardware store on an Indian Reservation has hired a local painting contractor to repaint the inside of the building. The only time that the painting contractor has access to the facility is during business hours. It is assumed by the manager that the contractor knows and follows all public safety regulations. It is brought to the manager's attention that the contractor has been sanding and painting the ceiling over the produce area without providing any protection to the produce. Once the manager investigates the complaint she notices that paint dust has fallen on the produce, resulting in a white film on much of it.

    a.   How should the manager handle the customer's complaint?
    b.   How could this problem have been prevented?
    c.   Should anything be done if your tribe has no regulations or codes to handle this situation, but that outside the reservation the use of this particular paint is known to be a health problem?
    d.   What should the manager do with the produce covered with the white film?
    e.   Who would be liable if a customer became ill after consuming some of the affected produce?

12. *Native American Issues Problem 12*. (Native American, Black Hills, Sioux Nations, South Dakota). [lrh]. In 1868 the Ft. Laramie Treaty confined the Oglala Lakota, one tribe of the great Sioux Nation, to the western half of present day South Dakota in exchange for promises of rations, annuities, agencies, schools, physicians, blacksmiths, teachers, etc., to be provided over an undetermined number of years. Today the reservation's size has decreased from over 40 million acres to an area of 50 by 90 miles. According to a U.S. Supreme Court ruling in 1980 the Sioux Nation was awarded $125 million for the loss of the Black Hills. To the Oglala Lakota, Paha Sapa (Black Hills) is Wakan (holy, sacred) and cannot be bought or sold at any price. The Oglala Lakota have refused to accept the $125 million. Instead, they want Paha Sapa returned to them as it was before the U.S. government illegally seized it in 1877. Today no one knows what has happened to the $125 million that the U.S. Supreme Court awarded, none of which has gone to any of the Sioux Tribes.

    a.   Do you feel that the Black Hills of South Dakota should be returned to the great Sioux Nations?
    b.   Discuss why it is very likely that the Black Hills will remain in the control of non-Native

12. *Native American Issues Problem 12*. [lrh] (continued)

    Americans regardless of the past injustice to a minority.

    c. If the Black Hills are returned to the Great Sioux Nation, how might this affect the economy of the state of South Dakota and the U.S. government?

13. *Native American Issues Problem 13*. (Potentially Responsible Party, Retroactive Liability). [rt]. In the 1930s and 1940s a uranium mining company hired 350 Navajos to mine high-grade uranium ore in the northern portion of the Navajo Reservation. During this time period, safety regulations and health risks were neither explained nor made available to the workers. Recently, many of the former employees, their children, and grandchildren are showing signs and symptoms of radiation exposure (cancer, loss of hair, genetic abnormalities, etc.). When these individuals attempted to file lawsuits against the responsible parties, they were met with roadblocks: companies no longer in business or bought out by other companies, no records showing names of former employees, etc.

Assuming you are a lawyer hired by the affected parties, answer the following questions.

    a. How would you proceed to determine who are the potentially responsible parties (PRPs)?

    b. How would you prove that the affected parties were former employees of the mining company that supposedly hired them?

    c. Since the mining operation took place so long ago, can the mining company really be held responsible?

14. *Native American Issues Problem 14*. (sustainable agriculture, restoration, Navajo reservation). [rt]. In the southwest United States lies a vast reservation of 17.5 million acres known as the Navajo Nation. The boundary of the Navajo Nation stretches across northwest New Mexico, northeast Arizona, southeast Utah and the southwestern-most corner of Colorado. The topography of the reservation varies from low elevation desert regions to high elevation mountainous regions.

Modern anthropological theory describes the Navajo people (Dine) as semi-nomadic, having ventured throughout the southwest before settling in their present location. They quickly adapted to the use of horses and other livestock introduced into the region by the Spanish. Currently the Navajo subsist primarily on herding sheep. Many areas of this region have been overgrazed by

14. *Native American Issues Problem 14.* [rt] (continued)

livestock, and have reverted, through a series of ecological transitions, toward increasingly less desirable grasses and shrubs, and unstable or unprotected surface soils. Each ecological recession has bred more erosion; and the erosion has bred a further recession of the vegetation. The result has been a progressive deterioration of the region's ecosystem involving not only the plants and soils, but the animal communities as well. There are often days when dust storms occur which block out the sunlight and make breathing difficult.

a. Assume that you are a professional engineer. How would you work with the people of the Navajo Nation to begin to resolve or reverse what is an enormous air pollution problem?
b. What sustainable agriculture and development approaches would you recommend and why?
c. How long would you predict it would take for the proposed sustainable solution approach to show improvement in either the air quality or in the ecological health of the region?
d. Which agencies (tribal, state, or federal) would you need to work with to solve this problem?
e. What are some of the factors you would need to consider in order to reach a workable solution using a sustainable approach?

15. *Native American Issues Problems 15.* (base closure, Navajo Tribe, Zuni Tribe, defense property conversion). [rt, rrd]. A former Army munitions storage site (Fort Wingate) is located just east of Gallup, New Mexico. Recent court battles between the U.S. Army and the Bureau of Indian Affairs, representing the Navajo and Zuni Tribes, have resulted in a ruling which will eventually give the land back to these two tribes. What measures should be taken to ensure the safety and well-being of any tribal member that might eventually occupy this land, and which agencies would be involved?

16. *Native American Issues Problems 16.* (BIA, BLM, dipping vats). [rt]. During the 1930s to 1940s sheep ticks and sheep scabies became a major problem for the Navajo livestock producers on the Navajo Reservation. To address this problem, officials from the BIA and the BLM constructed cement-lined trenches (dipping vats) that were approximately 12 feet long by 3 feet wide by 4 feet deep. A series of steps at one end of the vats allowed the animals to walk into and out of the vats and to then drip-dry in a fenced area. The fenced area contained a concrete floor which sloped downward toward the vat so that as little dip solution as possible would be lost

16. *Native American Issues Problems 16.* [rt] (continued)

by seepage into the ground. It is believed that the dip solution used in these vats contained DDT. Although the dipping vats are no longer used, they have been declared a health hazard by the Navajo Nation EPA. The tribe's budget has little money that can be allocated towards the cleanup of this problem. Suggest some ways in which the Navajo Tribe can remedy this health risk.

17. *Native American Issues Problem 17.* (Native Americans, environmental justice, Southwest Lawrence Trafficway, development, wetlands protection, environmental impact statement, NEPA, Haskell Indian Nations University). [teb].

Background Information
Lawrence, Kansas, a community with a population (1996 estimate) of approximately 89,900 and home to the University of Kansas and Haskell Indian Nations University, has for some time experienced significant, on-going congestion along certain traffic corridors. One of the primary areas of congestion occurs along 23rd Street, a major east-west corridor having extensive commercial frontage. This corridor also serves as the route connecting US Highway 59 to Kansas Highway 10. To alleviate this congestion, construction of a controversial trafficway along the southern and western boundaries of the community is underway. Although the idea of a route which would loop around the city to its south has existed since the 1930s, it has only been since 1985 that the issues and controversy surrounding such a trafficway have become more contentious.

In June of 1985, the Douglas County Commission announced that a $3.5 million bond issue was being considered for the purpose of constructing a 14-mile long trafficway that would connect Kansas Highway 10 east of the city with the Kansas Turnpike on the northwest. On November 6, 1990, and following several heated public hearings and a failed lawsuit attempting to block the issuing of bonds without a public vote, county residents voted 13,679 to 10,815 to support $4 million in bonds for the trafficway. The western portion of the trafficway, a 9-mile leg connecting the Kansas Turnpike on the northwest to US Highway 59 at 38th Street on the city's south edge, was opened for traffic in the fall of 1996. Figure 17 shows a map published in the *Lawrence Journal-World* (November 21, 1996) announcing the opening of what is considered the western or first leg of the South Lawrence Trafficway (SLT). At the time of this opening, a decision on where the final leg alignment of the trafficway would go had not yet been made and this decision was surrounded by conflict.

Nearly a year later the final leg of the trafficway to complete the connection between the Kansas Turnpike and Kansas Highway

17. *Native American Issues Problems 17.* [teb] (continued)

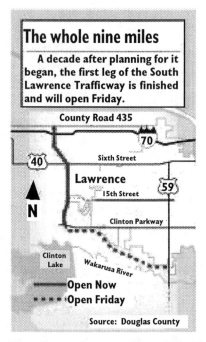

The whole nine miles

A decade after planning for it began, the first leg of the South Lawrence Trafficway is finished and will open Friday.

County Road 435

70

40   Sixth Street

Lawrence   59

15th Street

Clinton Parkway

Clinton Lake   *Wakarusa River*

N

━━━ **Open Now**
▪ ▪ ▪ ▪ **Open Friday**

Source: Douglas County

Terry Stevens/Lawrence Journal-World

Figure 17. Western portion of the Lawrence Trafficway, Lawrence, Kansas.

10 was effectively halted by a U.S. District Court injunction. On December 4, 1996 the Douglas County Commission voted 2-to-1 for selecting the trafficway's final leg alignment to be constructed within the existing 31st Street easement (see Figure 18). The decision came after considering three possible alignments. Based on this analysis, the 31st Street alignment was believed to be the most cost-effective option, and the only one which could obtain the necessary regulatory approval to be built.

As shown in Figure 18, the alignment selected would cross the southern boundary of the Haskell Indian Nations University campus. American Indian supporters and environmentalists (KU Environs and its president, and the Wetlands Preservation Organization, a coalition that includes the Haskell Indian Nations University Student Senate) filed suit in the U.S. District Court on March 12, 1997. The suit requested that the progress toward completing the eastern leg of the trafficway be stopped until a supplemental environmental impact statement (SEIS) could be completed. The concern was that by not waiting for the SEIS to be completed, the impact of the trafficway on the spiritual ceremonies and prayer conducted on the south Haskell campus, on Baker

17. *Native American Issues Problems 17.* [teb] (continued)

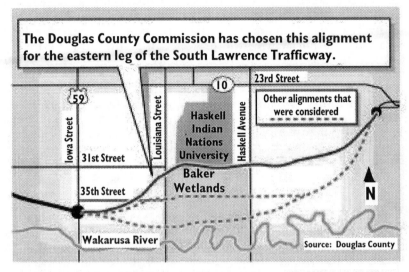

The Douglas County Commission has chosen this alignment for the eastern leg of the South Lawrence Trafficway.

23rd Street

Other alignments that were considered

Haskell Indian Nations University

Baker Wetlands

31st Street

35th Street

Wakarusa River

Iowa Street

Louisiana Street

Haskell Avenue

N

Source: Douglas County

Figure 18. Selected and alternate routes for eastern portion of the Lawrence Trafficway alignment, Lawrence, Kansas.

Wetlands areas, as well as any direct impact to other wetlands were being ignored. The U.S. District Court issued a permanent injunction on July 17, 1997 preventing any further action or expenditure of money on the final leg until the SEIS could be completed.

Complicating Issues

Federal agencies having a role in the issues and debate surrounding the SLT have included the U.S. EPA, the Federal Highway Administration (FHA), and the U.S. Army Corps of Engineers. The U.S. EPA has maintained that a final alignment, and in particular the 31st Street alignment, should not be selected until the SEIS is completed and the impact on Haskell evaluated. The primary influences on EPA's position regarding the trafficway's alignment have been a presidential order on environmental justice and the interest in allowing the National Environmental Policy Act (NEPA) process to be completed. Although EPA is obligated to review and comment on the SLT project because of NEPA, EPA does not have any authority to slow or stop its progress.

The FHA, also opposed to the proposed 31st alignment, announced on December 6, 1997 that it was withdrawing its involvement with the SLT project. In the announcement the FHA agreed that the $10.4 million of federal funds approved and appropriated for the project could be considered as having been spent on the first 9 miles of the trafficway. This decision effectively

17. *Native American Issues Problems 17.* [teb] (continued)

ceased any further obligation for FHA involvement and has left the
U.S. Army Corps of Engineers as the only remaining federal agency
having any authority over the project.   In addition to the FHA's
decision, the state's transportation department confirmed that no
additional federal funds would be requested to complete the final
leg of the trafficway.   The U.S. Army Corps of Engineers, having
the responsibility of regulating construction in federally protected
wetlands, issued a construction permit in 1993 for the 31st Street
alignment prior to the presidential order on environmental justice.
While the Corps has begun looking into the issues and concerns
that have been expressed by Haskell, it is not clear how these
concerns would affect the conditions of the existing permit.

Questions
Note: The following represents only a short list of possible
questions or topics that could be considered relevant to the issues
surrounding the SLT. The instructor is encouraged to use these as
a basis for leading open discussions or in designing a role-play
exercise culminating in a mock public meeting or hearing.

a.   What is the presidential order on environmental
justice and what does it focus on?

b.   How might environmental justice apply to
Haskell Indian Nations University and the
South Lawrence Trafficway?

c.   What is the purpose of the National Environmental Policy Act (NEPA)?

d.   What is an environmental impact statement
(EIS) and what does it address?

e.   List and describe the primary environmental
impact issues which you feel an EIS concerning
the South Lawrence Trafficway should be
addressing.

f.   Under what circumstances might a Supplemental EIS be necessary?

g.   When was the executive order on the protection
of wetlands signed and what legislation was it
intended to enhance?

h.   Under what provision would the Wetlands
Protection Executive Order apply to the South
Lawrence Trafficway?

i.   Explain the role and responsibility of the U.S.
Army Corps of Engineers regarding wetlands
and the protection of wetlands.

j.   What are engineering options that might be
considered for minimizing the potential impacts
to either Haskell or the Baker wetlands during

17. *Native American Issues Problems 17.* [teb] (continued)

the design and construction of the final leg of the trafficway. Include drawings of any design elements or concepts proposed. Complete this for each of the three alignment alternatives: 31st Street alignment, the 35th Street alignment and the southern-most alignment.

k. Are there any benefits that might be realized by Haskell if the 31st Street alignment were completed?

III. SOLUTIONS

1. *Native American Issues Solution 1.* [lrh].

   a. The Pine Ridge Indian Reservation, location in southwestern South Dakota, was where this leak occurred. The community was Pine Ridge, South Dakota.

   b. Possible solution:
      • Sample and test the soil surrounding the leaking UST site for petroleum contaminants using the appropriate sampling and analytical methods.
      • Install groundwater monitoring wells around the site and determine the direction of groundwater flow.
      • Collect groundwater samples to determine the extent and depth of free product, if any, and the size and possible movement of the contaminate plume.
      • Review the station's product inventory records (purchasing and sales) and if possible determine the length of time the tank has been leaking and the gross volume of product lost.

   c. There are bacteria naturally occurring in soil that are capable of breaking down hydrocarbons. Factors which could limit these bacteria from degrading the hydrocarbons associated with this leak are the lack of an appropriate electron acceptor (oxygen) or an inadequate availability of nutrient (e.g., nitrogen or phosphorous) in the soil which is needed to stimulate the aerobic degradation of the contaminants. In order to ensure that both of these factors are provided, the contaminated soil could be excavated and transported to a land treatment facility (soil farm). A land treatment facility is a site having specifically designed treatment beds or units that are constructed to accept contaminated soil and prevent the leaching of hydrocarbons, either into the soil beneath the site or in the storm water runoff from the site. In these treatment units or beds, a high nitrogen-content fertilizer is added to the contaminated soil, and the soil is mixed and aerated by tilling. In most instances, soil moisture content must also be managed.

1. *Native American Issues Solution 1*. [lrh] (continued)

>These operations stimulate microbial activity and effectively speed the biodegradation process. Over time the hydrocarbons are degraded. Although the time required for complete soil remediation using microbial degradation will depend on the proper management of soil aeration and nutrient addition, as well as other factors, new remediation techniques are continually being developed which may enable these degradation times to be shortened.

2. *Native American Issues Solution 2*. [lrh]. The answer will vary depending upon the tribe selected and its cultural traditions, the depth to which the student is willing to study these cultural traditions, and the how these cultural traditions are interpreted.

3. *Native American Issues Solution 3*. [lrh]. These are open-ended problems designed to provide the student with an opportunity view the environment through another culture's perspective. Although it should be attempted to be answered from a Native American viewpoint, a Euro-American viewpoint would also be acceptable. The questions posed could also be used as the basis for either a research or debate activity. The answer presented below is only one example of responses that might be provided to these questions.

>a. Yes. Concerned Native Americans living on reservations can initiate public environmental education programs to develop greater awareness of those activities which can adversely impact the environment or impair the reservation's natural resources for future generations. Additionally when financial resources allow, tribes can invest in the education of environmental professionals who can assist the tribe with the range of environmental issues and related problems that the tribe will become involved with.
>
>b. Native Americans will need to develop a clear understanding of the scope of environmental problems they face from both a cultural and technical perspective. Skills needed will include, but are not limited to: a greater awareness of tribal cultural traditions; training in environmental professional and technological areas (chemical, environmental, or sanitary

3. *Native American Issues Solution 3*. [lrh] (continued)

> engineering; environmental chemistry, law, politics, and science, etc.); and greater networking and communication with other tribes on environmental issues.

4. *Native American Issues Solution 4*. [lrh]. Native American tribes are under the jurisdiction of the U.S. EPA. Tribes can develop their own codes and regulations according to U.S. EPA guidelines. (Note: This answer can be found in EPA's rules and regulations documents as Part 501-State Sludge Management Program Regulations.)

5. *Native American Issues Solution 5*. [lrh]. The following discussion provides one example of how Native American tribes are dealt with by the U.S. EPA on environmental-related issues. Under the eligibility requirements for primacy for sludge management, Native American tribes are considered by U.S. EPA to be equivalent to states.

> (A) Consistent with Section 518(e) of the CWA, 33 U.S.C. 1377(e), the Regional Administrator will treat an Indian Tribe as eligible to apply for sludge management program authority if it meets the following criteria:
> (1) The Indian Tribe is recognized by the Secretary of the Interior.
> (2) The Indian Tribe has a governing body carrying out substantial governmental duties and powers.
> (3) The functions to be exercised by the Indian Tribe pertain to the management and protection of water resources which are held by an Indian Tribe, held by the United States in trust for the Indians, held by a member of an Indian Tribe if such property interest is subject to a trust restriction on alienation, or otherwise within the borders of an Indian reservation.
> (4) The Indian Tribe is reasonably expected to be capable, in the Regional Administrator's judgment, of carrying out the functions to be exercised, in a manner consistent with the terms and purposes of the Act and applicable regulations, of an effective sludge management program.
> (B) An Indian Tribe which the Regional Administrator determines meets the criteria described in paragraph (a) of this section must

5. *Native American Issues Solution 5.* [lrh] (continued)

> also satisfy state program requirements described in this part for assumption of the state's program.

6. *Native American Issues Solution 6.* [lrh]. Native American tribes are protected from State laws due to the treaties with the United States government. The U.S. EPA has authority over Native American tribes for environmental regulations. However, this authority can be considered questionable due to the Bureau of Indian Affairs' (BIA's) trust responsibility toward Native American tribes.

7. *Native American Issues Solution 7.* [lrh]. These are open-ended questions having the following as one set of possible solutions.

> a. The issue could be presented to the Tribal Council in an attempt to have the land committee reverse the decision concerning the use of insecticides, or at least delay its use. If time permitted, the location of the insectary might be moved (an expensive solution).
> b. A lawyer could be contacted to explore the legal options that might be available to recover investment costs. Additionally, the impact of the planned insecticide use could be publicized to potentially gain the support of the local public.

8. *Native American Issues Solution 8.* [lrh]. Responses to these questions will vary depending upon the significance the student places on ancestors and on burial grounds. This problem may be used as the basis for an in-class discussion and can provide an opportunity to view this issue from different perspectives.

9. *Native American Issues Solution 9.* [teb]. It can be expected that the environmental consultant would recommend to the tribe that the burial ground be relocated. Although this solution would require locating a new burial site and still delay the opening of the landfill and its operation, the delay experienced would not be as extensive as that which would result from the tribe initiating a search for a new landfill site. Depending on the value which a particular tribe places upon its ancestors and the sacred significance of a newly discovered burial site, the controversy and conflict this solution would invoke might range from little to no controversy or conflict to a strong resistance to even considering disturbing ancestral grounds.

10. *Native American Issues Solution 10*. [lrh]. The intent of this problem is to have students conduct enough background research on zeolite, mining practices, and southwest South Dakota so that the discussion of information needed is specific to this problem. At a minimum, students should determine what zeolite is, what health risks are associated with zeolite and its mining and handling practices, the environmental impacts resulting from zeolite mining, the impact of noise on nearby cultural ceremonial activities, the impact or loss of recreational or geological study areas, and the long-term profitability of the mining operation balanced with the impact on or loss of tribal grazing lands.

11. *Native American Issues Solution 11*. [lrh]. Although this is a somewhat open-ended problem, there should be some consistency in response such that an attempt to remedy the problem is made and that a concern for the potential health impact on the store's customers is considered. Additionally, it is suggested that this problem be used as the basis for an in-class, role-play exercise. Variations of this exercise could place students in the role of a Native American or non-Native American playing either the manager, customer or contractor. Change the make-up of these roles to allow students to view the problem from different perspectives and to explore possible differences in the outcome of the response to the problem. A role-play exercise will also provide the students with some experience in verbal communication and responding to public concerns.

One possible solution:

   a. The manager should talk with the customer about the problem and assure him that it will be discussed with the contractor and the problem will be remedied. The solution to this problem could continue depending on the customer's attitude.
   b. The manager could have provided the contractor access to the store during non-business hours or temporarily closed the produce area while it was being painted, and relocated the produce. This could be done for other areas throughout the store as well.
   c. The potential for such a problem should have been considered before the contractor began painting. Public health concerns should be a priority for all retail businesses. At a minimum, actions should be taken as soon as the potential for a problem is realized.
   d. This response should be based on the problem statement that outside the reservation the use

11. *Native American Issues Solution 11*. [lrh] (continued)

       of this paint is a known health problem. The produce should be removed and disposed of properly. It could be possible that the disposal of the produce would require that it be handled as if it were waste paint.

e.   With no laws existing on the reservation against using this paint or the painting practice used by the contractor, liability may be difficult or even impossible to establish. However, the manager should have an ethical and moral responsibility to the store's customers.

12. *Native American Issues Solution 12*. [lrh]. All of these questions should be considered open-ended and thus various responses are to be expected. The following is one possible response from the viewpoint of a Native American.

a.   Yes, an injustice was done. If our laws are to be valued they should be abided by and injustices should be corrected.

b.   It is very unlikely that the Black Hills would be returned, and in particular in its original condition before seized. Currently, the Black Hills exists as a recreational area used by the world's population, and mining activities are still on-going. The revenues from both the tourism and mining industries are substantial enough that the incentive for non-Native Americans to maintain control of the Black Hills is far greater than for returning it to the Sioux.

c.   The state of South Dakota would lose those revenues currently generated from tourism and mining activities occurring in the Black Hills region. Although arguable, the state government of South Dakota might not be able to exist as it currently does. A thorough response to this question should include a discussion of the current status of South Dakota's economy and government. The U.S. government would also be impacted due to the loss of mining revenues.

13. *Native American Issues Solution 13*. [lrh]. This problem is related to a real and on-going issue which has yet to be resolved in the courts. The responses to these questions should reflect and be based on standard practices and legal procedures that are currently in place or have had a precedence established and applied to abandoned sites. Additionally, possible solutions to these questions

13. *Native American Issues Solution 13.* [lrh] (continued)

may be based on information found in either **Theodore, M. K. and Theodore, L.**, *Major Environmental Issues Facing the 21st Century*, Prentice Hall, 1996, or **Holmes, G., Singh, B., and Theodore L.**, *Handbook of Environmental Management and Technology*, Wiley Interscience, 1993.

14. *Native American Issues Solution 14.* [lrh]. The problem presented has yet to be resolved and due to the vastness of the reservation, it is not anticipated that a solution will be achieved in the near future. The responses to the questions posed, however, should reflect current methods or practices in sustainable agriculture or ecological restoration. It should also be expected that the response would include discussion of newly proposed concepts and methods that might be applicable. Various tribal, state and federal agencies would be involved, as well as researchers at various universities familiar both with the region's many ecological areas and with the problem.

A list of some, but not all of the factors that would need to be considered in developing a sustainable agricultural and development approach for the area includes: the species originally found in the various ecological areas; soil types; meteorological characteristics; fugitive dust emissions, transport, and ambient air concentrations; successful and unsuccessful sustainable approaches used in similar settings; and cost-benefit analysis of implementing sustainable methods and ecological restoration.

15. *Native American Issues Solution 15.* [rt, rrd]. The Base Realignment and Closures Branch of the Department of Defense has the primary responsibility for transferring DoD properties to the private sector. As this area was a former Army munitions storage site, concerns should be raised regarding unexploded ordnance left behind, chemical waste sites and spill areas associated with weapons storage and maintenance areas, and other waste sites (USTs, landfills, etc.) that would exist on the base in support of the personnel that were stationed there throughout its history. Site assessments and remedial investigations would be conducted as part of the base's environmental management plan. Supported by the outcome of this baseline monitoring, the removal and remediation of contaminated locations would have to be carried out. Once completed, these remedial activities would generally remove contaminant to levels below which human exposure would be limited or non-existent. The trend in remediation, however, is to reduce cost by allowing contaminant to remain in place at levels which are considered to produce acceptable risk. This is contrasted to historical remediation approaches which resulted in attempts to treat contaminated sites to background levels regardless of cost or resultant human or environmental exposure risks.

15. *Native American Issues Solution 15*. [rt, rrd] (continued)

With higher levels of contaminant being left in place on contaminated sites, the potential exposure to chemicals is increased. To ensure that no unacceptable exposures to tribal members occur in the future, all conditions and development limitations prescribed by deed restrictions must be adhered to. These could take the form of restrictions on well drilling and groundwater use, extent of subsurface excavation that is permitted for the construction of dwellings, etc.

During the process of base clean-up and closure, and base property transfer, the agencies involved could include: the U.S. Army, BIA, OSHA, U.S. EPA, and all parallel state agencies.

16. *Native American Issues Solution 16*. [rt]. Until the tribe is able to allocate enough money towards clean-up, the vats will likely remain in their current, contaminated condition for some time. An immediate action to remedy the health risk would be to insure that human and animal access to the vats is strictly restricted. Possible long-term solutions might include approaching any one or a combination of the following agencies: BIA, BLM, U.S. EPA, or the Arizona Department of Environmental Quality and request assistance for the physical removal of the vats and decontamination of soils surrounding the dipping area.

17. *Native American Issues Solution 17*. [teb]. This is an open-ended question. As indicated in the problem statement, the instructor is encouraged to use the questions posed as a basis for leading open discussions or in designing a role-play exercise culminating in a mock public meeting or hearing. Efforts should be made to explicitly identify all of the stakeholders in this debate; their objectives and motivations regarding the SLT; numerous similarities and differences in the cultural perspectives and motivations of the stakeholders; and advantages and disadvantages of each alternative from the perspective of each stakeholder group. Attempt to reach a consensus that maximizes the benefits to all stakeholders. Compare your outcome to that which resulted in the real-life case as reported on the current web page for the Lawrence Herald-Journal, Lawrence, Kansas.

Chapter 12

Ethics

Howard Beim

I. INTRODUCTION

The need to make ethical decisions does not start the day one enters his/her first job. Encounters with ethical concepts begin early in life. Sometime in our youth, before most of us have any active memory, our parents almost certainly encouraged us to tell the truth. The conflict between choosing the ethical act of telling the truth and accepting the punishment that may follow or choosing the unethical path of lying and going unpunished introduces virtually everyone to the struggle for ethical behavior that lasts a lifetime. In school, it is not long before we learn that it is unethical to copy homework and to cheat on tests. The unethical nature of stealing from a classmate is another area that surely received some attention early in our lives. As we grow up, the ethical situations often become more complex. Conflicts develop. Situations arise in which we may have to decide if it is more ethical to hide an unpleasant truth that will hurt someone or to stay loyal to the truth that our parents taught us to value so highly.

As we move into professional life, we learn long lists of ethical dos and don'ts. Some examples are:

- never falsify data
- never offer or accept a bribe
- judge colleagues, employees, employers, clients, and others on their professional merits and not on any other basis such as race, gender, religion, age, national origin, or disability
- never draw misleading conclusions from the data you have collected
- avoid conflicts of interest
- accept work only in those areas in which you are properly trained
- obey all laws, rules and regulations
- never break confidentiality
- always do the best possible work of which you are capable
- statements to the public must always be truthful and objective
- the health and safety of the public must always be the highest priority
- always act to protect the environment.

Ethics becomes a dilemma for the professional when confronted with difficult choices or apparent conflicts between two ethical imperatives. A modern term for the outcome of such dilemmas is "situational ethics" which refers to the temptation to bend the rules when a situation seems to have no satisfactory or satisfying solution.

Examples of difficult choices are:

- A choice between keeping one's job and the income that is needed to care for the children who depend upon you and refusing to certify substandard work as ordered by your employer (job versus family)
- A choice between whether or not to get involved when you become aware of unethical behavior on the part of one's colleagues or employer even though not personally touched by it (job versus conscience)

Conflicts between two ethical imperative situations could include:

- the need to observe the confidentiality of a client after learning of the client's unethical behavior (protection of the client versus unethical behavior), and
- the ethical imperative to faithfully serve an employer or client versus the ethical imperative to protect the public (protection of the employer versus the public). At what point does a professional stop working within the system and report a violation of regulations to the authorities?

The second conflict of ethical imperatives stated above introduces issues that an environmental management professional is likely to encounter. Most professionals belong to one or more professional organizations which commonly adopt a code of ethics for their members. Each code is designed to provide members with an ethical foundation from which to work. A look at what such codes say about this particular conflict is instructive.

The American Society of Civil Engineers code:
Engineers shall hold paramount the safety, health and welfare of the public in the performance of their professional duties. [This is known as CANON 1.]
Engineers who have knowledge or reason to believe that another person or firm may be in violation of any of the provisions of CANON 1 shall

present such information to the proper authority in writing and shall cooperate with the proper authority in furnishing such further information or assistance as may be required.

The National Society of Professional Engineers code:
Engineers shall hold paramount the safety, health and welfare of the public in the performance of their professional duties [II. RULES OF PRACTICE].
Engineers shall at all times recognize that their primary obligation is to protect the safety, health, property and welfare of the public. If their professional judgment is overruled under circumstances where the safety, health, property or welfare of the public are endangered, they shall notify their employer or client and such authority as may be appropriate.

IEEE code:
We, the members of the IEEE ... commit ourselves to the highest ethical and professional conduct and agree:
1. to accept responsibility in making engineering decisions consistent with the safety, health and welfare of the public, and to disclose promptly factors that might endanger the public or the environment

It is clear that these codes place the ethic of protecting the public and the environment over the ethic of faithful service to the client or employer.

Variations in circumstances can produce an infinite number of dilemmas arising from conflicts and difficult choices. One needs to approach each case by reducing the problem to its essentials and then applying the rules and principles that have been agreed upon as professionals to the root cause of the ethical question. That may not make action any easier, but should make the ethical decision a clearer one to choose.

II. PROBLEMS

1. *Ethics Problem 1*. (ethics, values). [cmd]. How are ethical values, by individuals and by a society, determined?

2. *Ethics Problem 2*. (utilitarianism, duties ethics, rights ethics, virtue ethics). [cmd]. It has been generally accepted despite cultural variations that any historical ethic can be found to focus on one of four different underlying moral concepts. Identify and define each of these.

3. *Ethics Problem 3*. (community). [cmd]. Aldo Leopold made the following observation on personal ethics in his 1949 *A Sand County Almanac*, "...The scope of one's ethics is determined by the inclusiveness of the community with which one identifies oneself." In terms of the concept of environmental ethics, how does the concept of "community" expand beyond that captured in the cited concept of personal ethics?

4. *Ethics Problem 4*. (development, preservation, U.S. Forest Service). [cmd]. If a consortium of investors from California and Germany was proposing to build a ski resort on U.S. Forest Service land near your property and, as a rule, you like to ski, would you be likely to use the facilities? If so, do you think this is "evidence" that you prefer development of public lands over forest resource preservation? Do citizens always "vote with their pocketbooks?" Why?

5. *Ethics Problem 5*. (government agencies, profit, multiple-use, U.S. Forest Service). [cmd]. The U.S. Forest Service has a "multiple-use" directive, i.e., the lands and resources it administers provide a variety of goods and services to the public. For example, this means that the land and resources provide not only recreational opportunities but revenues from timber harvests and sales as well. Should government agencies such as the U.S. Forest Service strive to make a profit? Why or why not? How do you think your ethics affect your opinion on this issue?

6. *Ethics Problem 6*. (environmental ethics, conservation, engineering ethics). [hb]. As an engineering student, how do you incorporate environmental awareness into your everyday ethics?

7. *Ethics Problem 7*. (ISO 14000, codes of ethics, audits). [hb]. ISO 14000 is a voluntary standard for environmental management systems. A company can declare itself in conformity with the standard, but third-party certification of conformity is available and is generally regarded more highly by purchasers. The company seeking third-party certification contacts a "registrar" who sends an

7. *Ethics Problem 7.* [hb] (continued)

audit team of three trained professionals to review the company's management system. The audit usually takes 3 days.

On the third day of an audit of a facility, one member of the team reveals to the lead auditor that he has worked for the facility recently as a consultant. Is this an ethical problem? What should the lead auditor do?

8. *Ethics Problem 8.* (environmental management styles, attitudes). [tb, hb]. Based on the results of a survey (Petulla, J. M., Environmental Management in Industry, in *Ethics and Risk Management in Engineering*, Albert Flores, Ed., University Press of America, Lanham, Maryland, 1989), companies may be placed into one of three categories with respect to the attitudes of their managers towards environmental laws and hazardous waste disposal.

The first category, represented by 29% of those surveyed, is defined as having what is called "crisis-oriented environmental management." Companies in this group generally will have no full-time environmental manager or other personnel assigned to manage their environmental concerns, will provide minimal resources for environmental considerations, and will tend to fight environmental regulations. Management of these companies typically consider it cheaper to pay fines and lobby than to proactively work on environmental compliance.

The second category, represented by 58% of those surveyed, is considered to have "cost-oriented environmental management" and will establish policies on environmental matters as well as separate units for managing environmental matters for the company. Although these companies accept regulations as being a business cost, they do not do so with much enthusiasm nor commitment.

The third category, represented by 9% of those surveyed, has what is referred to as "enlightened environmental management," and is responsive to environmental concerns. These companies have administrative support in staffing environmental divisions, acquire and use state-of-the-art equipment, and maintain good relationships with governmental regulators.

You are a young environmental engineer and have just completed an interview for a new environmental management position with a relatively small company that manufactures metal specialty products. Until now, this position has been non-existent within the company. During your interview, the company's president openly praised the company's "responsible attitude toward the environment," citing an example of employees organizing the recycling of aluminum and other metals from certain lathing operations, as well as the company's "clean record" with the state regulatory agency.

8. *Ethics Problem 8.* [teb, hb] (continued)

While touring the facility's metal degreasing process area, you notice two employees through an open doorway in the rear of the building pouring liquids from two 5-gallon waste solvent containers directly onto the ground. In the area where the liquids are being poured the ground is extremely discolored and void of vegetation. The vegetation nearby is visibly stressed. You have been offered the position at an attractive salary rate and are asked to make a decision within 2 weeks.

    a. How would you place this company within Petulla's environmental management categories?

    b. What would be your decision regarding employment with this company?

    c. If you were to take the job, what would be your approach to the environmental management attitude you witnessed during the interview? Would you report the improper discharge of solvents to the local regulatory authority? Support your decision from either a professional or personal ethical standpoint, or both, and be explicit.

9. *Ethics Problem 9.* (ISO 14000, codes of ethics, audits). [hb]. Ralph has a Ph.D. in environmental engineering, a Professional Engineer (P.E.) License, and a job as an assistant professor at a university. To learn more about ISO 14000 and to earn some extra money, Ralph became a certified ISO 14000 auditor. During an audit of a facility, Ralph observed a significant violation of a U.S. EPA regulation. ISO 14000 is a voluntary standard of environmental management systems. It is not a regulation. ISO 14000 audits are confidential.

What does Ralph do? Does he forget about the significant violation since he is not a U.S. EPA inspector and it does not violate the ISO 14000 standard he is there to audit? Does he insist it be part of the audit report? Does he call it to the attention of the facility management? Does he report it to the U.S. EPA?

10. *Ethics Problem 10.* (incinerators, NIMBY). [hb]. The passage of a number of environmental laws in recent times, coupled with educational campaigns on the part of environmental groups and a supportive citizenry, have resulted in a substantial reduction in the quantity of hazardous waste produced. It is not likely, however, that the production of hazardous waste can be completely eliminated. As long as hazardous waste is generated, it will be necessary to dispose of that waste as safely as possible.

10. *Ethics Problem 10.* [hb] (continued)

Incineration is an effective method of hazardous waste disposal and it has become safer than ever with the passage of the Clean Air Act of 1990. Yet, local communities (often with the help of vocal environmental groups) have so successfully opposed construction of hazardous waste incinerators in their areas that no new facilities have been built in the U.S. in many years.

Discuss the ethics of opposing the construction of a hazardous waste incinerator in your local community. Propose a solution to the problems raised by such opposition.

11. *Ethics Problem 11.* (landfills, hazardous waste, realtors, environmental liability). [hb]. The following account was taken from an article by R. A. Stallings, Social Movements and Dramatic Events: Closing a Toxic Waste Landfill, *Journal of Hazardous Materials*, 27, 27-48, 1991:

BKK Landfill is the largest commercial hazardous waste management facility in California. It opened in the early 1960s as a Class I disposal site receiving, with few exceptions, the full range of industrial and household waste. The facility, which has a total area of 583 acres (2.36 km²), is located on a hilltop in the City of West Covina. This landfill was the only operating facility of its type within a 150 mile (240 km) radius of Los Angeles. Disposal of municipal waste and sewage treatment sludge took place on 157 acres (0.64 km²) of the site while another 92 acres (0.37 km²), partially encircled by the refuse disposal area, was a receiving site for over 84 million gallons (3.2 x 10⁹ L) of hazardous liquids. It was estimated that the landfill would reach its full capacity by the early 1990s.

At the time that BKK Landfill started its operation, the surrounding area was undeveloped and uninhabited. During the late 1960s and early 1970s, however, when housing demand increased substantially, new housing tracts were built along the base of the hillside at the eastern, western, and southern edges of the landfill. By the mid-1980s, nearly 40,000 residents lived within a 1 mile radius from the center of the landfill. The new suburban residents then encountered the problem of having a toxic waste site in their backyards.

Public awareness of possible health hazards increased as the residential community grew larger. The Not-In-My-Back-Yard (NIMBY) attitude toward the toxic waste disposal facility became widespread. A number of social activists began to oppose the operation of the landfill. Opposition activities included a petition drive to close the facility in 1980, formation of several active protest groups, a local ballot initiative to close the landfill and picketing of the entrance to the toxic waste facility in 1981. By 1983, the

11. *Ethics Problem 11.* [hb] (continued)

Department of Health Services (DHS) had been requested to hold many public hearings to determine what health hazards had been generated by BKK Landfill. In early 1984, the congressional representative for West Covina demanded that U.S. EPA terminate the operation of the toxic waste disposal site at once.

Aside from the instinctive disapproval of having the landfill nearby, the neighborhood was unhappy over numerous trash spill incidents, truck accidents on the major transportation lines leading to the entrance of the landfill, and occasional foul odors. The greatest concern, however, was the health risk posed by emissions of carcinogens. In 1982, vinyl chloride was detected in one of the new housing developments along the southern border of the facility. High readings of trichloroethylene (TCE), perchloroethylene (PCE), and ethyl dichloride were also detected in 1983, and methane gas was detected in adjacent property in 1984.

An inspection by U.S. EPA in 1984 found that there was underground migration of the liquid waste and its by-products in every direction out of the facility. Additional inspections discovered the existence of fractured bedrock revealing that underground liquids could not be held as originally claimed. A bowl of sandstone was also found underneath the toxic waste site instead of impermeable material. The contaminated groundwater migrated downward since the southern underground barrier was constructed on a bed of fractured shale. The State DHS then admitted that its earlier geologic studies of the disposal site were in error. The U.S. EPA and the California DHS prohibited receipt of further liquid wastes and approved BKK Landfill plans to dig more wells to prevent contaminated groundwater from leaking through the sandstone barriers.

Several risk assessment studies were conducted by the University of Southern California's medical school in 1980 and 1983 to investigate the incidence of cancer in the adjacent community. Both studies showed neither health hazard to the surrounding residents, nor statistically significant increases in the rate of cancers in the tracts adjacent to the landfill. In spite of these findings, high emissions of vinyl chloride and leaks of contaminated groundwater resulted in a long-term evacuation effort. The evacuation lasted a total of 7 months (from July 1984 to January 1985) partly because some evacuees refused to return home after their homes were declared safe. The evacuation caused many serious, unexpected consequences. The evacuated homes were burglarized, forcing the landfill to arrange a security patrol at exceedingly high cost. Other concerns of the residents were the loss of property values. In spite of its drastically reduced income, BKK Landfill had to pay for all the accommodations for the evacuees and a large sum for fines from several lawsuits. As a result, BKK

11. *Ethics Problem 11.* [hb] (continued)

Landfill publicly announced that it would no longer dispose of toxic material after August 31, 1984 but would continue to receive non-toxic solid waste. Yet, it was still unable to satisfy the public. The BKK Landfill closure plan was rejected. Several requirements were added by government agencies. Site stabilization, planning, financing, and continuous gas monitoring are examples of the added requirements. The facility was then asked to provide complete closure of the landfill within 10 years.

Discuss the ethical behavior, or lack of it, on the part of each of the following parties in the above true story: the landfill operators, the land developers and realtors, the residents, and the government agencies. Try to put yourself in each of their places. What lessons could be learned from this bit of history?

12. *Ethics Problem 12.* (ethics, hazardous wastes, politics, transportation). [dj]. Consider the following hypothetical scenario where three adjoining states have conflicting needs for disposal of hazardous wastes.

The citizens of New Jersey (population of 12 million, an urban state with little available land) are reluctant to dispose of their hazardous waste in their own state and are forbidden under federal law to dump it at sea. They have the political will and the financial ability to dispose their waste out of state. To do this, they must transport their waste through the state of Pennsylvania, and store it there temporarily until completion of a treatment, storage and disposal facility (TSDF) planned for construction in the state of West Virginia. About half the waste is currently accumulating under conditions that are not considered safe for storage beyond a period of 10 years. Land in New Jersey is either too scarce, too expensive, or too close to densely populated areas for a TSDF, and hazardous waste incineration was voted down as an option 2 years ago.

The state of Pennsylvania (population 16 million) has decided to dispose of its hazardous wastes in several engineered landfills that are permitted to contain properly containerized hazardous wastes. The landfills were sized to allow for a 50-year operating life before closure based on projected hazardous waste generation rates within the state. New Jersey's wastes would be transported along a corridor that would expose about 1 million additional people to risks of transportation accidents, and then temporarily stored north of a large city in the western part of the state. Pennsylvania does not want New Jersey's hazardous waste temporarily stored on its soil and has threatened to pass legislation banning the temporary facility. It also has threatened to sue New Jersey in federal court if the latter tries to contract for out-of-state disposal of its wastes that involves interim storage in Pennsylvania.

12. *Ethics Problem 12.* [dj] (continued)

The state of West Virginia (population 3 million) is relatively poor and rural. Its economy has been devastated by layoffs from steel mills and coal mines, and its citizens and elected officials are very interested in attracting new industries to the state. A TSDF capable of accepting wastes from several states in the region, including New Jersey's, would produce several hundred jobs during the construction and operation phases and would be a potential source of tax revenue to the impoverished state. The largely blue-collar labor pool is used to industrial hazards and is receptive to the opportunity to be retrained to treat, store and dispose of container-ized wastes. There is widespread support for the construction of a regional facility that would dispose of both West Virginia's and New Jersey's wastes. There is a small vocal opposition that is concerned about the environmental risks and does not want West Virginia to be perceived as a dumping ground for other states. The governor of West Virginia has contacted his influential U.S. senator, who is holding hostage a federal water quality bill that funds projects for water quality improvements that are administered jointly with Pennsylvania until Pennsylvania becomes more compliant on the hazardous waste transportation and interim storage issue.

The federal government, in the persons of the Congress, various regulatory agencies, and the courts may be called upon to participate in the resolution of the dispute among these states.

There are several political rationales that could be used to resolve this dispute:

1. A "greatest good for the greatest number" rationale could be applied, i.e., reducing the risk to the 12 million citizens of New Jersey outweighs the increased risk to the 1 million potentially exposed citizens of Pennsylvania.
2. Alternatively, a "minority rights" rationale could be applied, where large populations are not allowed to worsen the quality of life of smaller populations. In this case, the minority is numerical, not ethnic.
3. An "each takes care of their own" rationale could be applied, where each state must locate and develop, no matter what the cost, a hazardous waste TSDF within its own boundaries. This rationale deprives West Virginia of economic benefits, reduces risks for Pennsylvania, and increases disposal costs for the taxpayers of New Jersey.

List the major risks associated with:

12. *Ethics Problem 12.* [dj] (continued)

    a. Allowing hazardous wastes to remain in New Jersey under partially unsafe conditions in a scenario where New Jersey "loses" its bid to transport its hazardous waste out of state.

    b. Transporting New Jersey's hazardous waste via Pennsylvania (with interim storage in Pennsylvania) under a scenario where New Jersey "wins" its dispute. In formulating the answer, consider the increase in volume of hazardous waste transportation and storage that would occur in Pennsylvania.

    c. Disposing of New Jersey's hazardous wastes at a RCRA-permitted facility in West Virginia.

    d. How can New Jersey reduce the risks associated with transporting its hazardous waste through Pennsylvania?

13. *Ethics Problem 13.* (ethics, hazardous wastes, politics, transportation). [dj]. Consider the rationales described in Problem 12 above.

    a. List the pluses and minuses, or "winners" and "losers," for each of the rationales described in the problem statement.

    b. Each rationale above is cast in "win-lose" terms. Are there alternative rationales that are "win-win" or "lose-lose?" For example, under Rationale #1, could New Jersey compensate Pennsylvania for the increased risk New Jersey exposes Pennsylvania to? The payments could be used for road, rail or river improvements and increased surveillance of hazardous waste shipments. What would be involved in determining a fair price for this compensation?

    c. For a class project, find and summarize some case studies in the library that involve actual disputes among state governments. A classic example is the recent interest of the Mescalero Apache Reservation in operating a nuclear waste repository over the objections of some of the other citizens of New Mexico (including the governor and the legislature). In reading the literature on a particular dispute, try to identify:
        • the nature of the disposal problem (type of waste and disposal method), and

13. *Ethics Problem 13.* [dj] (continued)

- the motivating interests of the parties in conflict, i.e., who stands to gain from having the proposed facility and why, and who stands to be hurt by the facility and why.

III. SOLUTIONS

1. *Ethics Solution 1*. [cmd]. Without a doubt, culture has a strong bearing on the formation of ethical standards by individuals and by a society. The beliefs, knowledge and traditions that make up a culture provide society with a world view that frames the range of acceptable behavior for any given individual and for the society as a whole. It is essential to recognize, however, that the source of ethical standards must be external to the individual, whether grounded in religious faith, social tradition, or professional or governmental policy. An individual who claims for him or herself the right to establish the ethical rules by which they live can just as easily disestablish those rules should they subsequently prove inconvenient to observe.

2. *Ethics Solution 2*. [cmd]. The four different underlying concepts which influence ethical development are:

- Utilitarianism - which focuses on actions which lead to the greatest good for the greatest number of people.
- Duties ethics - which are composed of certain principles of responsibility or duty, even if the greatest good does not result.
- Rights ethics - which emphasize the rights of people affected by an act.
- Virtue ethics - which focuses on acts that result in an individual becoming a morally "good" person.

3. *Ethics Solution 3*. [cmd]. Answers to this question can vary depending upon the respondent's definition of the concept of environmental ethics. The answers could include a personal ethics definition of community as those *humans* with whom one deals on a daily or almost daily basis. "Community" within the concept of environmental ethics can include elements of the non-human environment also, such as land, water, air, animals, plants, etc.

4. *Ethics Solution 4*. [cmd]. This open-ended question will be answered according to the preferences (ethics) of the respondent. If the respondent says he/she would not use the facility, it might be because he/she does not want such a facility, with its high volume of traffic and other "loss of natural quality" concerns, to be located near his/her property. There might be concerns about the removal of revenue such a facility would generate from the local economy. Personal philosophy might just be against such a narrow use of public lands.

If the respondent says he/she might use such a facility, the respondent might believe such a use for public lands is an

4. *Ethics Solution 4.* [cmd] (continued)

acceptable definition of "multiple use." Furthermore, the respondent might feel that this use is appropriate at that particular location and would welcome the proximity of a high-quality recreational facility. The respondent may additionally believe that such a use would enhance his/her property value.

While "economic benefits" drive most individuals' decision-making, in the case of public lands, "voting with one's pocketbook" may not prevail. While the construction of the ski resort might increase the value of the nearby properties, the marring of the nearby pristine natural area might decrease the overall quality of life for residents of the area. These are personal choices.

5. *Ethics Solution 5.* [cmd]. This open-ended question will be answered according to the opinion, based on ethics, of the respondent. If the respondent feels that it is all right for the U.S. Forest Service to make a profit, ethical considerations might be based on the "mission" of the Forest Service - that of "multiple-use" in which the resources are managed for the production and sale of timber resources. If the resources are going to be harvested and sold, then getting the most profit form their sale makes good economic sense.

If the respondent does not believe the agencies of the federal government, like the Forest Service, should strive to make a profit, such views might come from the belief that a federal agency should not compete with private entities. The respondent may also be totally against the harvesting and sale of any resources from public lands and that they should be left in as pristine condition as possible for the non-consumptive use and enjoyment of the public at large.

6. *Ethics Solution 6.* [sl]. Firstly, we should realize that human habitants are just one group of many diverse habitant groups that comprise the ecosystem. Secondly, we must understand that we need to keep the ecosystem intact. Abuse of the environment might ultimately lead to catastrophe. Finally, we should keep in mind that the natural resources on this planet are limited, so we should all do our share of recycling and conserving energy.

7. *Ethics Solution 7.* [hb]. It is an ethical problem for an auditor to have a relationship with the company being audited. It is considered to be a conflict of interest in that the auditor may wish to please the company with a good report so as to enhance his/her chances of working with that company in the future.

The lead auditor must:
- call the ISO registrar and inform the registrar of what has happened,

7. *Ethics Solution 7*. [hb] (continued)

- remove the compromised member of the team,
- void all of the work done by the removed team member,
- request the company to permit an extension of the audit, and
- redo all of the work done by the removed team member.

In the event the redone work confirms the findings of the disqualified auditor, the registrar should apprise the auditor of this fact in writing. While the removed auditor clearly exhibited poor judgment in accepting the appointment to the team, his/her reputation should not remain compromised in the absence of incriminating evidence.

While it is to be expected that an engineer would recognize so clear-cut a conflict of interest as described, the problem, except for deliberate fraud, could be obviated by the simple expedient of having each auditor screened by a pre-appointment questionnaire.

8. *Ethics Solution 8*. [hb].

   a. Mixed messages were sent during the interview process as to exactly which category this company fits into. Consider the following factors:

- The environmental manager position is a new staff position within the company which would imply that, in the past, minimal attention was given to environmental considerations. This would put the company in the first "crisis-oriented" category.
- The company's president does seem to be interested in encouraging greater environ-mental awareness, and the new position reflects this. Having a person whose job is dedicated to environmental matters would put the company into the second "cost-oriented" category.
- At the least, there appears to be a definite problem with the employees' understanding of what it is to be "environmentally responsible." Environmental management consists of much more than just recycling. The discharge of solvents on the back lot indicates that environmental management is not being

8. *Ethics Solution 8.* [hb] (continued)

taken seriously. It appears that the company is in the first "crisis-oriented" category but moving towards the second "cost-oriented" category.

b. Only you could say if you would take the job, but remember this: a committed environmentalist who accepts this position may succeed in gently and progressively nudging the company into full environmental responsibility. If such an applicant withdraws, leaving the responsibility to other engineers of lesser integrity, unwilling to stand on principle, compliance will less likely be achieved. If you take the job and succeed environmentally, the excitement, the sense of achievement, and the ethical rewards may far outweigh the potential risk to employment security. You are the only one who knows best what you are looking for in a job...and in a life.

c. Although only the reader could supply the answer as to how he/she would handle the job, codes of ethics provide some guidelines. Most codes call for professionals to protect people and the environment as well as to act in the best interests of their client or employer with honesty, and integrity. Generally, this means that every effort must be made to work within the company to educate it to behave in an environmentally responsible way including compliance with the regulations. If that fails, ethics requires the professional to report the problem to the regulatory authorities. Such action could cost the professional his/her job. The codes do not provide escape clauses to allow the professional to keep his/her job at the cost of people or the environment.

9. *Ethics Solution 9.* [hb]. The question of professional ethics presented here can be generalized as follows: does a professional have an ethical duty to report any violation of which he/she becomes aware? Most licensing organizations have codes of ethics to which their professionals must subscribe. These codes generally call upon the professional to serve their client/employer with loyalty and to protect the welfare of the public. The codes, however, are sufficiently vague so as to not provide a clear-cut answer to Ralph's situation. In the absence of such a clear-cut code, there is no one right answer and your answer may vary from the one presented below.

9. *Ethics Solution 9.* [hb] (continued)

Ralph must first investigate further to make certain of his facts. If the violation is confirmed and has never been reported, he must insist that it be part of the findings presented to management at the exit meeting and in the report to the ISO Registrar who makes the final decision on awarding certification. Although ISO 14000 does not require compliance, knowingly failing to comply could affect the evaluation of the facility's commitment to good environmental management system practices that the standard does require.

Having done all that, does Ralph go the last step and report it to the U.S. EPA? Such a report would violate the confidentiality of the audit and might even precipitate a lawsuit. Ralph is also concerned that he may never be hired again as an auditor. He senses a conflict between two sets of ethics: the ethics of a professional bound to protect the public, and the ethics of a professional bound to confidentiality. Keeping silent would be easier but ethics is not easy.

It would be easy to answer "report it!" but the real question is what would YOU do?

10. *Ethics Solution 10.* [hb]. NOTE. This is an open-ended question with no single correct answer. Your answer may have addressed points that were not made here, and vice versa.

On the issue of opposition to the construction of the hazardous waste incinerator, it is not unethical to protect one's self, family, and community from a risk. This is especially true concerning the risk from, for example, a hazardous waste incinerator, over which one has so little control. Even with the best design and pollution controls, no one can absolutely guarantee that the incinerator will have zero emissions. Nor can anyone absolutely guarantee that the low level of emissions will not pose a hazard to anyone. Although the risk may be estimated to be as low as one excess cancer or other disease in each one million exposed people, this is no consolation if that one additional affected party is you or a member of your family.

It would, however, be unethical to avoid your risk by shifting it to someone else. The manufacture and use of products, such as furniture, rugs, clothing, cars, and electronics, have become an integral part of our way of life. Wastes are produced by the manufacture of all these products. As long as the products continue to be used, there is a moral obligation to share the risk that results from the disposal of the waste generated in making these products.

Sometimes, widespread opposition to hazardous waste disposal results in industry finding creative ways to eliminate the production of a waste that they cannot dispose of. Those who are successful in advocating the Not-In-My-Backyard (NIMBY) policy can then claim the highest of ethical behavior since all society had

10. *Ethics Solution 10.* [hb] (continued)

benefited from the improved production processes. All too often, unfortunately, industry simply moves its operations to a state or country where environmental restrictions are less severe. In addition to possibly shifting the risk to someone else, the NIMBY policy also causes the harm of loss of jobs. Since one cannot ascertain whether opposition to the hazardous waste incinerator will lead to the elimination of the production of the waste or the shifting of the risk to someone else, this behavior cannot be classified as ethical.

On the issue of problems raised by such opposition, a possible (but not perfect) solution may be a national hazardous waste incinerator siting program. Each region would be required to accept its share of incinerators. A government agency, or some other neutral body, could evaluate all potential sites in each region based on factors such as availability of facilities, accessibility, number of people affected, etc. A lottery system could be used to make the final choice among the top, roughly equal sites. To mitigate, somewhat, the risk imposed on the selected site, benefits, such as free health care or reduced taxes, could be provided to the people in the selected area. In this way, the risks and alternatives would have to be squarely faced by virtually everyone. It would not be possible to simply evade the problem by shifting it to someone else.

11. *Ethics Solution 11.* [hb]. This is an open-ended question with no one correct answer. Your answer may be different from the one given below.

The story, as told, does not reveal unethical behavior on the part of the landfill operators. The area was uninhabited when the site was built. The State DHS erred in its geologic study of the site. The landfill operators installed monitoring devices when leaks were discovered, paid for people to evacuate and to protect their homes in their absence. In the end, complete closure was proposed. All of these actions are ethical and proper. It is possible that the BKK Landfill operators knew more than the story tells us and failed to take proper action to protect people and the environment. Based on the story, however, there is no such evidence upon which to accuse them of unethical behavior.

The land developers and realtors who sold the land may or may not have known of the hazards presented by the close proximity of the landfill. Ethical behavior (as opposed to legal behavior) required that they investigate the possibility and to inform potential buyers of any possible risks. The story is silent on that matter so it is not known if investigations were or were not done. It seems unlikely that so many people would have bought homes in that area had they been so informed. What is your opinion?

11. *Ethics Solution 11.* [hb] (continued)

The homeowners immediately attract our sympathy. Their actions at the beginning of the story reveal, however, not the highest ethical behavior. Imagine that you own an auto repair shop. It is a small shop and your sole source of a livelihood. You operate honestly, legally and possess all the right permits. No one lives next door to you on any side of your shop. One day someone builds a house on the lot next to your shop and insists that you close your shop and move away because the noise disturbs his young children. Would that be ethical?

Once the leaks are discovered, the homeowners have a right to insist that the landfill not allow the chemicals to escape the boundaries of the landfill. It would certainly be ethical to attempt to recover any damages caused by those leaking chemicals. Would it be ethical, however, to seek damages if no damages had been *proven*?

The government agencies certainly do not garner any accolades for their role of mismanagement in this story. Despite that, there is nothing in the story that reveals unethical behavior.

The lessons to be learned from this story are:

- Landfills are a poor method of disposal of hazardous liquids,
- Look very carefully at the whole area when you buy a home, and
- There is not always someone to blame when a story ends badly.

12. *Ethics Solution 12.* [dj].

    a. The major risks associated with allowing the wastes to remain in New Jersey include the following:

- If storage conditions deteriorate, then accidental releases of the hazardous wastes could occur. If they are non-volatile, then release will occur primarily to soil, and, if the wastes are liquid, and released in sufficient volume, to groundwater.
  Contaminated soil would pose a risk to persons who would use the site after the wastes are removed, especially if accidental or purposeful dermal, inhalation or ingestion could occur. Children

12. *Ethics Solution 12.* [dj].

or adults playing on or working the soil could have traces of hazardous chemicals on their skin, or, they could inhale soil dusts contaminated with the hazardous compounds, or, if crops were ever grown at the site, they could eat foods contaminated with traces of the hazardous materials. Also, young children may eat the soil directly.

Soil erosion by wind or water off the site could transport hazardous chemicals onto adjoining lands or into local water bodies, beginning a process of transport that could affect other persons downstream or downwind.

If hazardous materials reach the local groundwater table, then down-gradient transport could occur, gradually contaminating an aquifer that could be a source of potable drinking water.

It should be noted that the costs of cleaning up a hazardous chemical release are generally far higher than the costs of preventing hazardous chemical releases in the first place; so, storage conditions should not be allowed to deteriorate. Hazardous wastes at the site should be carefully segregated and should be stored in chemically inert, corrosion-resistant, fire-resistant, and impact-resistant containers.

- If the materials at the site continue to accumulate to unsafe levels, i.e., more materials than the facility was originally designed to handle, then the unsafe storage conditions could develop, where potentially reactive combinations of wastes are stored too close to each other. If not properly segregated and stored, then a small accident at the facility, such as a fire, could have disastrous consequences with potentially explosive or flammable combinations of chemicals entering into the chemical reaction, with consequences far worse than would occur if the compounds were properly segregated and stored.

12. *Ethics Solution 12.* [dj].

    b. The risks associated with transporting the hazardous materials to the temporary storage facility in Pennsylvania are:

- Risks of a hazardous material transportation accident in route from New Jersey to the site in western Pennsylvania. A hazardous waste spill could expose travelers and communities along the route to potentially harmful levels of the compounds. Transportation planners would need to assist the planning process for moving the wastes by selecting roads and travel times that would reduce the risk to the maximum extent possible. Also, the transported materials should be properly contained (in DOT-approved containers), manifested, labeled (on each container), and placarded (on each truck or rail car). Some transportation requirements are in RCRA rules (40 CFR), but others are in Department of Transportation regulations (60 CFR).
  If New Jersey's wastes are to be transported along approved hazardous waste routes where other wastes are already being transported, then planners must consider the increase in risk associated with higher volumes of hazardous material traffic along those routes.
  In general, it would be most preferable to select transportation routes that are distant from population centers, and remote from schools, hospitals, nursing homes and other populations that are not easily removed from the site of an accident. Congested roads and intersections that have high accident rates should also be avoided.
- Risks of a materials handling accident during the process of loading the materials onto trucks or rail cars in New Jersey, or during off-loading of the hazardous materials from the trucks or rail cars at the temporary site in

12. *Ethics Solution 12.* [dj].

western Pennsylvania. Precautions should be taken to make sure that the personnel who transfer the wastes into approved storage containers, and who load the wastes onto and off of the trucks or rail cars, are properly trained in the handling of these materials, and that they exercise due care in the handling of the materials during the transfer operations.

c. The risks associated with disposing of New Jersey's wastes at a RCRA-permitted facility in western Pennsylvania are:

- Risk of accidental release during off-loading of the materials from trucks or rail cars. Precautions should be taken to make sure that the personnel who transfer the wastes off the trucks or rail cars are properly trained in the handling of these materials, and that they exercise due care in the handling of the materials during the transfer operation. Also, materials should be transferred to RCRA-approved interim or final storage containers.

- Risk of accidental release during processing of the wastes for final disposal. For example, inadequate combustion temperature or time in a thermal oxidizer could lead to excessive waste concentrations in stack gases or in waste ash.
  If wastes are to be chemically stabilized (by immobilization in a pozzolanic cement or in glass), then the stabilized waste forms must be tested for resistance to leaching of the immobilized wastes when exposed to water.

- Risk of release of any still-hazardous residuals that are placed in a landfill for final disposal. The landfill must be designed to minimize risk of potential releases. This usually means a primary containment system, consisting of geo-membranes and clay liners, then a secondary containment and leachate collection system, an impermeable cap

12. *Ethics Solution 12.* [dj].

that minimizes access of water to the stored wastes, and, finally a set of monitoring wells that would permit the detection of any contaminant plume that might reach groundwater.

The site that is selected for final disposal of the wastes must be geologically suitable for disposal. This means that the site should not be subject to faulting or soil swelling from any cause, and should ideally be underlain by soils and rocks that are of low permeability. Groundwater should be sufficiently below the lowest part of the site so that it will never rise up into the storage zone, and if there is a risk of this happening, then a pumping system should be engaged to dewater the site to prevent leakage of groundwater into the zone where wastes are stored.

To detect accidental releases of hazardous wastes, a RCRA-permitted treatment, storage and disposal facility (TSDF) must be monitored during its operation, and also must be monitored for many years after closure of the site.

d. New Jersey can reduce the risks associated with transporting its wastes through Pennsylvania by:

- Minimizing the volume of waste that is actually transported. The fewer the trucks or rail cars, the lower the overall chances of an accident. This can be accomplished by reducing the volume of the wastes that are stored in New Jersey. Volume reduction can be carried out by:

  Exercising pollution prevention actions at waste generators. Waste generators should reuse and recycle as much material as possible before shipping the wastes to the storage facility in New Jersey. Also, *waste generators could engage in waste exchanges,* whereby the wastes generated by Company A are sold as raw materials to Company B.

12. *Ethics Solution 12.* [dj].

An example of this is the use of fly ash from power plants in the manufacture of cement and asphalt.

- Treating the wastes to reduce its volume and/or toxicity. This can be done by:

  Removing water or other easily separable liquids from the wastes via evaporation or distillation,

  Burning the wastes to generate energy, reduce toxicity, and/or reduce volume prior to shipment. Of course, this opens up the sticky question of installing an incinerator (also known as a thermal oxidizer) in New Jersey, an alternative that may not be politically or economically feasible,

  Separating hazardous components using precipitation, reverse osmosis, electro-dialysis, ion exchange, microfiltration, ultrafiltration, or nanofiltration.

- Exercising all feasible precautions during the handling and transportation of the wastes through Pennsylvania. As previously discussed, this includes:

  Proper training of personnel who handle the materials during transfer, and *proper containerization, and proper manifesting, labeling and placarding.* Also the *waste transporters should be licensed and bonded, with vehicles that are regularly inspected, properly maintained, and meet all applicable safety standards.*

  Selection of transportation modes (road, rail or water), transportation routes, and transportation scheduling (time of travel and volume of waste per shipment) that minimize the risk to exposed populations.

13. *Ethics Solution 13.* [dj].

a. For the *"greatest good for the greatest number"* rationale, the "winners" are the people of New Jersey who have their risk of exposure to hazardous wastes reduced by transporting the wastes out-of-state. Property values in certain

13. *Ethics Solution 13.* [dj].

parts of the state that will be restored once the wastes are removed may also increase, causing some landowners to be "winners." Also, any party who stands to profit from the decision to transport the wastes is a "winner," including rail and truck transportation companies, waste processors, geotechnical and environmental consulting firms, and the employees and stock holders of the firms who will transport and process these wastes. Some of these "winners" may reside in western Pennsylvania or in West Virginia, the temporary and final repositories for the wastes, respectively.

Under this rationale, the "losers" may be the people of Pennsylvania who may have marginally higher risks of exposure to hazardous wastes as a result of the decision to transport New Jersey's wastes through the state. Also, property values in the vicinity of the "temporary" storage facility in western Pennsylvania may decline due to a perception that these properties are at increased risk of becoming contaminated by the wastes.

Under a *"minority rights"* rationale, the "losers" are the people of New Jersey, who now must dispose of their wastes in their own state at a higher cost. Tax costs and the costs of doing business in the state might increase as a result. Some businesses may elect to leave the state, increasing unemployment in the state for some people.

Property values in the vicinity of current disposal sites in New Jersey could decline, or begin to decline as a result of a decision to retain all wastes at current sites in New Jersey.

Some persons and companies in New Jersey could be "winners," as they may benefit from the decision to retain and store the wastes in New Jersey by becoming employed in the business of handling, treating and storing the wastes in the state. Clearly, the people of Pennsylvania are "winners" because there will be no increase in risk associated with the transportation of New Jersey's wastes through the state, or with the operation of a temporary storage facility in the western part of the state.

13. *Ethics Solution 13.* [dj].

Some Pennsylvania residents and companies may be "losers" because they will not be employed to transport and store the wastes in Pennsylvania.

The situation in West Virginia will be mixed. The state will lose the economic benefit of treating and disposing of New Jersey's wastes, so some persons and firms may be "losers." However, the image of the state in the minds of the American public may be improved, and lands that might have been devoted to waste treatment and storage could be put to other productive uses, such as agriculture, industry, housing, or, if in remote areas, for tourism. So, in some sense, the people of the entire state could be "winners" if West Virginia doesn't develop a reputation as the East Coast's dumping ground for hazardous wastes, and instead develops an image as a clean, green place to live.

Under the *"each takes care of its own"* rationale, the main advantage may be reduced political friction between the states. The influential West Virginia senator need no longer hold "hostage" legislation that could benefit Pennsylvania, and New Jersey, and Pennsylvania would not have to be at loggerheads over transportation routes and storage sites in their state.

However, a state with strained resources and limited space, such as New Jersey, may also "lose" somewhat, because their hazardous waste disposal facilities might be closer to sensitive populations. Also, if space is limited, a treatment, storage and disposal facility (TSDF) in New Jersey may not be as well-designed or risk-free as a TSDF operating in western Pennsyl-vania or in West Virginia.

Some economic sectors in the receiving states (western Pennsylvania and West Virginia) may "lose" because they will not be employed to transport, treat, store and dispose of New Jersey's wastes.

Real-estate holders in western Pennsylvania and West Virginia may "win" because their land will not decline in value as a result of the absence of any TSDFs for New Jersey's wastes.

13. *Ethics Solution 13.* [dj].

> However, some property holders, who could profit from the sale of their lands to firms that will operate the TSDFs, would "lose" because no sale would be transacted.
>
> b. Yes, there are alternatives to "win-lose" scenarios. If the political climate in Pennsylvania is not unduly hostile to waste transportation, then New Jersey could offer to compensate Pennsylvania for the increased costs associated with managing an increased volume of hazardous wastes traveling through the state. Payments could be made to enhance the training of emergency response personnel, or to improve road or rail links that would be used to transport the wastes so that accident risks are reduced.
>
> A "fair price" for this compensation might have to be worked out from the prior experiences of other regions that have dealt with this problem. If there is no region that has this prior experience, then reasonable cost estimates would have to be developed with the inputs of the several agencies that are involved.
>
> If the economic merits of operating the temporary facility in western Pennsylvania are apparent, then that part of Pennsylvania might "win" just as much, by benefiting from increased employment and an improved tax base, as New Jersey wins by ridding itself of the wastes.
>
> Implementation of any "win-win" scenario would require a lot of mutual trust, where all responsible parties adhere to their responsibilities in any agreements or contracts.
>
> c. The solution to this part of the question will vary based on the case study selected for analysis. The main focus of the analysis for any case study, however, should be an assessment of the motivating factors driving the various players in the dispute, and the ethics by which they are operating.

Chapter 13

ENVIRONMENTAL ACCOUNTING AND LIABILITY

Carolyn Daugherty

I. INTRODUCTION

One of the most significant challenges that has emerged in the 1990s for environmental managers and for those seeking employment as environmental managers is the anti-regulatory climate characterizing the attitudes of the U.S. Congress and state legislatures. With efforts to cut expenditures resulting in programs seriously underfunded at all levels of government, most state and local agencies are finding it increasingly difficult to maintain satisfactory levels of service mandated in federal and state laws designed to protect the health, safety, and welfare of citizens.

Nowhere is this budgetary crunch being more intensely felt than in the field of environmental regulatory enforcement which has traditionally kept most companies engaged in environmentally-sensitive activities on the "straight and narrow." As the government's ability to maintain and enforce regulatory standards wanes, who will step into this void of responsibility? Interestingly, it may turn out to be the companies themselves. The same companies that may have previously resisted governmental regulatory efforts now find it to be in their best interests to perform in an environmentally responsible manner.

Assisting in this corporate "reformation" have been the activities of other, non-governmental entities that have come to be known as "surrogate regulators:" banks, insurance companies, accounting firms, the stockholders and investors of these entities, and international standard-setting organizations. Because these surrogate regulators are largely concerned with corporate profitability, activities which adversely impact corporate environmental liabilities (such as the need for asbestos removal, improper storage of environmentally hazardous materials, and industrial processes which could pollute the environment) and which can result in drastic financial impacts on the corporation are being highly scrutinized by these new "regulators."

Each of these surrogate regulators' interests in environmentally responsible behavior on the part of corporations involves the identification of the sources of potential past and present liability, the financial responsibility incurred as a result of the liability, and the extent of the impacts which the environmental liability could impose on the corporation's profits. Concern for these issues by the surrogate regulators is being felt by Corporate

America, and companies with potential environmental liabilities wishing to remain profitable will become proactive in addressing current and future environmental liabilities.

Historically, employment opportunities for individuals trained in various environmental fields were largely to be found in the regulatory agencies of state and the federal governments. Employment opportunities by these agencies are also shrinking due to shrinking federal and state budgets for regulatory and environmental management activities. Today's opportunities in the environmental management field may be greater with the surrogate regulators due to their expanding role in environmental "regulation." This means that additional skills such as accounting, financial assessment, and insurance underwriting may be useful, if not required, by those seeking employment in the environmental field. Below is a brief discussion of the roles played by the surrogate regulators in environmental assessment and regulation enforcement and the specific opportunities with these entities where environmental managers may make the most significant contributions.

A. ACCOUNTING FIRMS

The interest of an accounting firm in the issue of a corporation's environmental liabilities stems from the extent to which the corporation is responsible for remediation of an environmental liability. The responsibility for remediation directly affects the relationship between a corporation's assets and liabilities, i.e., its profitability, and hence, the income which the corporation receives and extends to its shareholders.

The extent to which a corporation is involved in the creation of the environmental liability (solely responsible or in conjunction with other companies, known as potentially responsible parties, PRPs) and the ultimate cost of the remediation activities incurred by the corporation itself must be assessed and reflected as diminished profitability in their balance sheets.

Such activities are usually beyond the scope of traditional accounting and have led many accounting firms to engage the services, either in-house or contractually, with an environmental auditor. Clean-up costs for existing pollution are highly variable because every oil spill, asbestos removal, or landfill remediation project presents a different set of conditions that are often highly site-specific. An individual trained in the environmental sciences or environmental engineering can identify the various aspects of the problem that affects the cost of site remediation, and can offer meaningful estimates of clean up costs.

The role of accounting firms in quantifying the cost of environmental liabilities is still evolving. The field offers many challenges for environmental managers.

B. INSURANCE COMPANIES

Insurance companies have become surrogate regulators because corporations have become legally liable not only for current, individual pollution events, but also for the effects of pollution over long periods of time. Corporations that perform poorly in their environmental practices may be subject to higher insurance premiums or may not be able to obtain liability insurance at all.

The interest of insurers is with environmental impairment liability (EIL), which includes the effects of a company's processes and/or products on employees and users through time. While occupying a very small percentage of the insurance market, EIL insurance provides environmental insurance coverage and risk management benefits to all aspects of the marketplace, ranging from coverage of liability from the effects of poorly performed environmental services and construction, to hazardous waste transportation and site remediation.

For those trained in the environmental sciences and engineering, employment opportunities with insurance companies exist because such firms need sound risk quantification. This requires going out on-site and performing environmental audits and site surveys. According to one industry expert, what makes the EIL industry effective is quality underwriting. Rather than having a staff of trained insurance agents trying to understand the environmental engineering aspects of a business, the successful insurance company employs environmental engineers that learn the insurance business.

C. BANKS

Although assessment of environmental liability has not been a customary concern for bankers, it has been evolving into a necessary one. Beginning in 1985 some very startling court decisions awakened lending institutions to the fact that as lenders of money to corporations engaged in processes and products from which environmental liabilities could emanate, they become PRPs, paying remediation costs even if these costs are greater than the property value, original loan amount, or trust value.

To properly understand their role in the management of risk, the FDIC recommends that banks have an environmental risk management program in place to evaluate the impact on the value of real property of the adverse effects of environmental liabilities. Several of the components of such a program afford those trained in the environmental sciences and engineering employment opportunities. Banks should employ personnel having the appropriate knowledge and experience to evaluate the environmental risk associated with a piece of property. In this evaluation, it must be determined whether there might be some impairment to the property that would reduce its value, possibly incurring liability for the lending institution as a PRP, or if there is an otherwise unrecognized liability that would affect the financial viability or credit-worthiness of the borrower.

D. INTERNATIONAL STANDARDS

The International Organization for Standardization, ISO, quartered in Geneva, Switzerland, has been at the forefront in developing standards for environmental performance. These standards describe corporate environmental practices which must be achieved in order to receive certification that their processes and products meet environmental quality standards of acceptance required by many corporations doing business in the international market. Such standards prompt corporations to perform in an "environmentally friendly" manner to be able to conduct business, unimpeded, in all areas of the world. Expertise in the ISO certification process is an expanding field of opportunity for those trained in the environmental sciences and engineering.

E. SUMMARY

With decreased budgetary support for environmental regulatory enforcement at both the federal and state levels, it could be assumed that corporations have become less fettered by environmental laws and regulations in the application of their processes and the development of their products. However, the historically governmental regulatory role has been taken over to an increasing degree by "surrogate regulators." These non-governmental entities, the most visible of whom are accounting firms, insurance companies, and banks, are concerned with a corporation's environmental practices as they can become responsible for the costs of environmental remediation and the liabilities associated with it if their clients are irresponsible.

Increasing employment opportunities for those trained in the environmental sciences can be found with these surrogate regulators. For the most part, the surrogate regulators recognize that learning the banking, accounting, and insurance aspects of environmental liabilities is easier to acquire than the technical knowledge associated with environmental assessment and remediation procedures.

F. ABOUT THE PROBLEMS IN THIS CHAPTER

The problems in this chapter are focused on exploring the nature and role of these surrogate regulators in environmental management activities of businesses in the U.S. The causes and conditions of liability, tools for the evaluation of a company's profitability, and estimates of the impact of environmental liability on this profitability are also covered. Finally, a variety of references are provided throughout the problems to help orient the reader to this growing field of environmental management.

II. PROBLEMS

1. *Environmental Accounting and Liability Problem 1.* ("surrogate regulators," corporations). [cmd]. Due to government cutbacks in funding, regulatory-driven programs for most federal agencies are feeling the pinch. With fewer and fewer enforcement personnel, many feel that corporate environmental behavior will take a turn for the worse. One author suggests that such is not the case in the following article: **Buonicore, A.**, Surrogate regulators: a 'powerful' driving force in the industry, *Environmental Management*, 4, 86, 1996.

The traditional sources of pressure on corporations to perform in an environmentally responsible manner have come from federal and state regulatory agencies. However, with governmental cuts in funding, the role of these regulators is being diminished. It would be reasonable to expect, then, an increase in anti-environmental activity by corporations. However, such has not been the case because of the increasingly important role of other entities whose impacts on the performance of corporations may be even more significant. These entities have come to be known as "surrogate regulators." Who are these entities?

2. *Environmental Accounting and Liability Problem 2.* (potential environmental liability, banks). [cmd]. Why are banks interested in determining the potential environmental liability of corporations that seek to borrow money from them? (See Problem 1 for reference article on "surrogate regulators.")

3. *Environmental Accounting and Liability Problem 3.* (due diligence process, environmental liabilities). [cmd]. To what does the term "due diligence process" refer? How would this process reveal any environmental liabilities a corporation may have? (See Problem 1 for reference article on "surrogate regulators.")

4. *Environmental Accounting and Liability Problem 4.* (insurance). [cmd]. From an insurance standpoint, why is it in a corporation's best interests to be environmentally conscious? (See Problem 1 for reference article on "surrogate regulators.")

5. *Environmental Accounting and Liability Problem 5.* (SEC, disclosure). [cmd]. The Securities and Exchange Commission (SEC) requires corporations to disclose their environmental liabilities for accounting purposes. What happens to a corporation if it fails to properly report a liability? (See Problem 1 for reference article on "surrogate regulators.")

6. *Environmental Accounting and Liability Problem 6.* (potentially responsible party, acronym). [cmd]. In an effort to help

6. *Environmental Accounting and Liability Problem 6.* [cmd] (continued)

companies better account for environmental remediation liabilities on their balance sheets, the American Institute of CPAs has issued a statement of position which provides guidance on accounting for the environmental liabilities of PRPs related to the recognition, measurement, display, and disclosure of the liabilities. To what does the acronym "PRP" refer?

Refer to the following article which discusses this statement of position: **Gill, F. R.**, Environmental remediation liabilities, *Journal of Accountancy*, 180(3), 81, 1995.

7. *Environmental Accounting and Liability Problem 7.* (environmental remediation liabilities). [cmd]. Briefly describe what the concept "environmental remediation liabilities" refers to. (See Problem 6 for reference article on "environmental remediation liabilities.")

8. *Environmental Accounting and Liability Problem 8.* (entities, sources of liability). [cmd]. Environmental remediation liabilities have accrued to entities (companies and individuals) from three sources. Identify these sources of liability. (See Problem 6 for reference article on "environmental remediation liabilities.")

9. *Environmental Accounting and Liability Problem 9.* (causes of liability, conditions of liability). [cmd]. What is the underlying cause of an environmental remediation liability and under what conditions does an entity become liable? (See Problem 6 for reference article on "environmental remediation liabilities.")

10. *Environmental Accounting and Liability Problem 10.* (PRPs, loss-recovery situation). [cmd]. A Superfund site will cost $62 million to remediate. It has been determined that five entities constitute the PRPs. Four entities agree to participate in the remediation effort; one entity refuses to participate. Determine the following (see Problem 6 for reference article on "environmental remediation liabilities."):

    a. What will each of the four participating entities pay to clean up the site?

    b. What will happen to the non-participating entity?

11. *Environmental Accounting and Liability Problem 11.* (global economy, ISO 14000, level playing field). [hb]. One of the more notable trends of the last two decades has been the migration of corporations from the U.S. to other countries, where labor is

11. *Environmental Accounting and Liability Problem 11.* [hb] (continued)

cheaper and regulations are less burdensome. While this approach may be cost effective for the company in question, such use of the global economy undermines the effectiveness of U.S. regulations in the area of environmental management. As long as manufacturers can gain an economic advantage by evading U.S. regulations, our efforts to create a clean environment will encounter broad resistance based on the conflict between the legislative concern over health issues and industry's concern over their profit margin.

Suggest a way to level the environmental management system playing field for companies operating in the U.S. with those operating in other countries.

12. *Environmental Accounting and Liability Problem 12.* (present worth, time value of money). [wrl, rrd]. The concept of the time value of money is an integral part of most capital investment decisions, and environmental-related investments are no exception. The time value (also called present value) concept simply states that a dollar today is worth more than a dollar tomorrow. The following problem demonstrates this concept.

An investor may invest $60,000 in either Option A or Option B. The return on the investment for each option is given Table 29. The investor wishes to earn the highest rate of return possible. What is the present value of each option? Assume end-of-year discounting at a rate of 10%.

Table 29
Return on Investment for Investment Options A and B

Year	Annual Income Option A	Annual Income Option B
1	$10,000	$10,000
2	$15,000	$10,000
3	$10,000	$15,000
4	$10,000	$15,000
5	$15,000	$10,000
Total	$60,000	$60,000

The present value formula is as follows:

$$PV = AI \frac{1}{(1 + i)^n}$$

where PV = present value of annual income for period n in $; AI = annual income in $; i = annual interest factor or discount rate as a decimal; and n = number of annualized periods.

13. *Environmental Accounting and Liability Problem 13.* (batteries, disposal, rechargeable, photovoltaics, economics). [glh, rrd]. A good example of an applied environmental accounting and liability problem is that of the economics of disposable versus rechargeable batteries used in domestic applications. Throw-away batteries represent a significant non-point source of heavy metal pollution, particularly mercury and cadmium, into landfills. The major use of such batteries historically was the Type D size used in flashlights, but portable radios and tape players have made Type AA batteries much more popular. Battery manufacturers will now accept the spent batteries for recycle and/or disposal in a safe manner. Most communities have one or more collection sites - the town library, schools, etc. - where the public can turn in this hazardous waste.

A better solution to the battery problem is to switch to rechargeable batteries which can be reused over and over, lasting for years. These do require a charging device at some initial cost and a convenient 120 volt outlet for recharging. An even better system, one which is sustainable, employs the sun to do the recharging to eliminate power company involvement altogether. Again, a special solar charger is a necessity as is, of course, a sunny day or two (Real Goods, 1992).

a. Calculate the annual cost for the use of rechargeable Nickel-Cadmium AA and compare it to the annual cost for conventional batteries. They are to be used in a portable radio which uses four such batteries. The rechargeable batteries are recharged every month with a solar battery charger, which is not designed to get wet, and which requires 2 to 4 days to recharge the batteries depending on the cloud cover.

b. What is the payback time, if there is one?

Data:

	Conventional AA	Rechargeable Ni-Cad AA
Number required:	4	8
Cost (each)	$0.89	$2.75
Rotation frequency:	Monthly	Monthly
Solar recharger cost:	0	$14
Lifetime:	1 mo[†]	5 yr (1,000 charges)[††]

† Depending on intensity of use.
†† Claimed by manufacturer.

Reference: **Real Goods**, *Real Goods News*, Spring Issue, 10, 1992. Can be purchased by writing to c/o Real Goods, 966 Mazzoni Street, Ukiah, CA 95482-3471.

14. *Environmental Accounting and Liability Solution 14.* (treatment, risk analysis, transportation, liability, regulation). [wrl]. In evaluating the economic impacts of pollution prevention strategies, costs of alternative treatment options must be considered. In some cases, the reduction of potential liability alone may be adequate to justify capital expenditures for pollution prevention. The estimation of the clean-up cost of a spill or release is also an important factor. The following problem demonstrates a practical method of estimating transportation liability.

Biguns Chemical Co. transports solid hazardous waste to a disposal site. On average, Biguns' hauling trucks carry 4 tons of waste/trip for a total of 32,000 tons/yr. In the event of a truck overturn, it can be assumed that 2 tons of the waste are spilled. DOT statistics indicate that 1 out of 4,000 waste hauling trucks overturns during an average trip. Industry studies and Chevron Corp.'s SMART (Save Money And Reduce Toxics) data indicate that cleanups resulting from transportation spills cost as much as $10,000 per ton. Calculate the total annual potential liability of producing and "disposing" of the waste in this matter. Express the answer in $/yr and $/ton.

15. *Environmental Accounting and Liability Problem 15.* (depreciation allowance, straight-line method). [cjk]. Define the straight-line method of analysis that is employed in calculating depreciation allowances.

16. *Environmental Accounting and Liability Problem 16.* (payout time, scrubber). [cjk, dj]. When considering the cost of air toxics control technologies it is often important to know how long it will take to recover the cost of an investment. The time required to recover this cost is called the payout time.

Define the commonly accepted formula for payout time and calculate the payout time for the following scenario: A plant manager spends $10,000 for new scrubbing packing for a tower that strips air toxics out of a gas stream. The manager decides to depreciate the equipment at $1,430/yr (7-year straight-line method depreciation) and estimates that the equipment will generate $1,500/yr in annual profit.

17. *Environmental Accounting and Liability Problem 17.* (rate of return, scrubber). [cjk, dj]. A plant manager must often decide if it is truly worthwhile to invest in new equipment. The fundamental question is, "Will the company earn more money per year by investing in this equipment than it would if the money was invested in an interest-bearing financial account?" The calculation that is used to make this decision is called the percent rate of return on investment which is carried out as follows:

17. *Environmental Accounting and Liability Problem 17.* [cjk, dj] (continued)

$$\text{Percent rate of return} = \frac{\text{Annual Profit}}{\text{Initial Investment Cost}} \times 100$$

Calculate the percent rate of return on investment for the plant manager's $10,000 investment described in Problem 16.

18. *Environmental Accounting and Liability Problem 18.* (replacement cost, return on investment). [cjk]. Three different scrubbers are available for the removal of a toxic contaminant from a gas stream. The service life is 10 years for each scrubber. Their capital and annual operating costs are as follows:

Table 30
Capital and Annual Operating Costs for Various Gas Scrubber Options

Scrubber	Initial Cost	Annual Operating Cost	Salvage Value in Year 10
A	$300,000	$50,000	0
B	$400,000	$35,000	0
C	$450,000	$25,000	0

Which is the most economical scrubber? Use a straight-line depreciation method in your calculations and the following equation for the calculation for the rate of return based on two investment options:

Return on investment for choice between two alternatives:

$$\text{ROI} = \frac{(d_1 - d_2) + (o_1 - o_2)}{(c_2 - c_1)} (100)$$

where d_1 = annual depreciation of first alternative, $/yr; $= (c_1 - r_1)/n_1$, with n_1 being expected lifetime in years; d_2 = annual depreciation of second alternative, $/yr, $= (c_2 - r_2)/n_2$, with n_2 being expected lifetime in years; r_1 = residual value of first alternative, $; r_2 = residual value of second alternative, $; o_1 = annual operation and maintenance cost, first alternative, $/yr; o_2 = annual operation and maintenance cost, second alternative, $/yr; c_1 = capital cost of first alternative, $; and c_2 = capital cost of second alternative, $.

19. *Environmental Accounting and Liability Problem 19.* (sampling, statistics, environmental accounting and liability). [fwk]. This problem is designed to be an interactive team exercise simulating the decision making process involved in contaminated-

19. *Environmental Accounting and Liability Problem 19.* [fwk]
(continued)

site assessment and remediation. Teams of three to five students should be formed to address the following problem.

There is a narrow strip of soil, 30 yards long, along the side of Farmer Mudd's driveway. Neighbors report that he dug a single hole, and disposed of his pesticide waste from time to time in that hole. Replicate samples taken on a nearby farm that did not have a pesticide disposal problem yielded soil concentrations of the pesticide of concern of 7.2, 6.2, 8.8, 5.7, 7.5 and 6.6 ppm. In places where there had been disposal of pesticide, the concentration could be as high as 90 ppm. No migration associated with groundwater movement is expected.

It is estimated that the concentration of pesticide will taper off to a negligible value at a distance of 2 to 3 yards on each side of the center of the hole. It will cost $3,000 to bring the drilling crew to the farm. For each sample collected, it will cost $25 for the sample bottle and preservative, $15 for shipping, and $160 for analysis.

Your team should choose the locations (in whole yard increments from 1 to 30) at which you will take your samples. Your instructor or assistant will then provide your team with the results of sampling those locations. Once you have seen the results, you may wish to take additional samples. This is acceptable, but will cost you an additional $3,000 for the drilling crew.

When you know enough about where the contaminant is located, you should specify the portion of the strip to be remediated, in tenths of a yard, at a cost of $4,000 per lineal yard. You should remediate any portion of the property with concentrations of the pesticide above 10 ppm. Portions missed will cost you a penalty of $6,000 per yard. Your goal is to conduct the sampling and provide remediation at the lowest total cost. The team achieving the lowest cost will receive the highest grade (or the next job if this had occurred in real life!).

III. SOLUTIONS

1. *Environmental Accounting and Liability Solution 1*. [cmd]. The "surrogate regulators" include banks, insurance companies, accounting firms, investors and shareholders, international standard-setting organizations, and the public-at-large.

2. *Environmental Accounting and Liability Solution 2*. [cmd]. If corporate assets, such as real estate, contain environmental liabilities, then the bank incurs that liability in granting the loan. This liability could easily exceed the value of the asset it holds as collateral.

3. *Environmental Accounting and Liability Solution 3*. [cmd]. As defined by law, the term "due diligence" refers to "such measures of prudence, activity, or assiduity, as is properly to be expected from, and ordinarily exercised by, a reasonable and prudent man under the particular circumstances; not measured by any absolute standard, but depending on the relative facts of the special case."
   If, for example, a corporation is seeking a loan from a bank to purchase a piece of property to expand its operations, the bank would be required by FDIC regulations to include checks for any environmental liabilities associated with the property or with the corporation seeking the loan. This check is required in order to be sure that neither the borrower nor the property has any potential environmental liabilities which could make the bank responsible for any environmental liabilities needing remediation.

4. *Environmental Accounting and Liability Solution 4*. [cmd]. Poor environmental performance may result in significantly higher insurance premiums or, if past performance shows a blatant disregard for environmentally responsible behavior, the inability to obtain any insurance at all.

5. *Environmental Accounting and Liability Solution 5*. [cmd]. In addition to the fines, penalties, and aggravation it can incur for failure to report a liability, the SEC can suspend registration of a corporation's stock.

6. *Environmental Accounting and Liability Solution 6*. [cmd]. "PRP" stands for "potentially responsible party." This refers to an entity that is potentially liable for the costs of cleaning up a site which has been contaminated due to the entity's actions.

7. *Environmental Accounting and Liability Solution 7*. [cmd]. "Environmental remediation liabilities" refers to the costs companies would incur from being identified as a PRP responsible for cleaning up a site which has been contaminated through their actions at the site, or by their association with a disposer or treater that has acted in a manner to contaminate the site.

8. *Environmental Accounting and Liability Solution 8.* [cmd]. The sources of liability for environmental remediation come from the Comprehensive Environmental Response, Compensation, and Liability Act (CERCLA); the corrective action provisions of the Resource Conservation and Recovery Act (RCRA); or analogous state and non-U.S. laws and regulations.

9. *Environmental Accounting and Liability Solution 9.* [cmd]. The underlying cause is the past or present ownership or operation of a site or the contribution or transportation of waste to a site where remedial action must be taken. Liability results from the probability that an entity will be held responsible for participating in the remediation process and that the liability must be reasonably estimable.

10. *Environmental Accounting and Liability Solution 10.* [cmd].

    a. A total of $15.5 million will be paid by each of the four participating PRPs ($12.4 million of its own share and $3.1 million more which is its one-fourth share of the non-participating PRP's share).

    b. A lawsuit by the four participating PRPs would have to be brought against the non-participating PRP to force its contribution to the costs of remediation.

11. *Environmental Accounting and Liability Solution 11.* [hb]. This is an open-ended question with more than one answer. Your answer may not be the same as the one provided below.

One way to level the playing field would be to somehow require companies operating in other countries to abide by the same environmental management regulations as companies operating in the U.S. The U.S., however, has no authority to require companies (even American companies) in other countries to follow its regulations. The U.S. government could (and should) attempt to make treaties or reach understandings with other governments to adopt uniform regulations. Such attempts to persuade other governments that adhere to uniform regulations are in the best interests of the health of their citizens; however, this would be a very difficult task because of the need in those countries for foreign investment and economic development. While the attempt is still worthwhile and should go forward, it must be viewed as a very long-term process.

A broad international agreement on environmental management systems is another approach that has some precedent in other areas. The International Maritime Organization (IMO) is

11. *Environmental Accounting and Liability Solution 11.* [hb] (continued)

an United Nations organization to which virtually the entire world belongs. The IMO issues regulations for the transportation of hazardous materials by ship. Through this organization, these regulations are now uniform throughout almost the entire world. Perhaps a similar approach to controlling the hazardous chemicals that threaten the environment could be successful. Again, although this is another goal worthy of pursuit by government, it must be considered to be a very long-term effort.

Another approach would be to levy an import tariff on products produced in violation of U.S. environmental regulations. The tariff would equal the savings that not following U.S. regulations generated. Aside from the political difficulties in passing such legislation, determining what the correct tariff should be would be a monumental task. The U.S.does not have the manpower nor funds to inspect every plant in every country (not to mention a lack of legal authority). The bureaucracy needed to track and compute the tariffs would be enormous.

The ISO 14000 environmental management standard presents a relatively simple way of accomplishing, to a large degree, a level playing field. Legislation could require that all products imported into the U.S. be produced by facilities possessing third party ISO 14000 standard certification. Products produced without certification would face stiff tariffs. Although ISO 14000 does not equate to the observance of all U.S. environmental regulations, it would very substantially improve environmental management practices and result in a giant step towards global environmental management uniformity. Perhaps the best aspect of this approach is that it would be accomplished without a work force of U.S. inspectors, without significant additional government bureaucracy, and virtually no tax dollar expenditures.

12. *Environmental Accounting and Liability Solution 12.* [wrl]. For Option A Year 1, the present value of the income is calculated as follows:

$$PV = \$10,000 \ \frac{1}{(1 + 0.10)^1} \ = \ \$9,091$$

For Option A Year 2 the present value of the income is:

$$PV = \$15,000 \ \frac{1}{(1 + 0.10)^2} \ = \ \$12,397$$

The results for both options, all years, are given in the Table 31.

12. *Environmental Accounting and Liability Solution 12.* [wrl] (continued)

Table 31
Return on Investment for Investment Options A and B Expressed in Present Value Terms

Year	Annual Income Option A	Present Value Option A	Annual Income Option B	Present Value Option B
1	$10,000	$9,091	$10,000	$9,091
2	$15,000	$12,397	$10,000	$8,264
3	$10,000	$7,513	$15,000	$11,270
4	$10,000	$6,830	$15,000	$10,245
5	$15,000	$9,314	$10,000	$6,209
Total	$60,000	$45,145	$60,000	$45,079

From Table 31 it can be seen that Option A yields a higher present value than Option B ($45,145 versus $45,079). Option A is the better investment even though both options earn the same total undiscounted income.

This type of approach can be used to effectively evaluate different environmental options and to conduct meaningful environmental accounting and liability comparisons.

13. *Environmental Accounting and Liability Solution 13.* [glh].

a. The annual cost for conventional AA batteries under the conditions stated will be:

(4 batteries/mo) (12 mo/yr) ($0.89/battery) = $42.72/yr

The annual cost for rechargeable Ni-Cad batteries for the first 5 years is:

$$\frac{[(8 \text{ batteries}) (\$2.75/\text{battery}) + \$14/\text{charger})]}{5 \text{ yr}} = \frac{\$36}{5 \text{ yr}} = \$7.20/\text{yr}$$

b. Payback of the original capital cost of $36 would occur in:

$$\frac{\$36}{(4 \text{ batteries/mo}) (\$0.89/\text{battery})} = 10.1 \text{ mo}$$

After 10 months, the solar-rechargeable system is free, at least until replacement is required. If the claim of 1,000 charges is accurate, then 5 yr is a conservative estimate. Obviously, if they last more than 5 yr, then the economics

13. *Environmental Accounting and Liability Solution 13.*
[glh] (continued)

improve. Assuming the batteries last for 1,000 charges and the charger lasts indefinitely (not unreasonable) then the cost for four "new" batteries becomes less than $0.04, or a penny each. The cost of conventional batteries will not change (assuming no fluctuation in price).

As this problem illustrates, it makes good economic and environmental sense to switch to solar rechargeable batteries. There will be some inconvenience due to the necessity of charging the spent batteries, but not unlike having to run to the store (more wasted energy) to buy new conventional batteries. Of course, if the battery charger is left out in the rain, you may need a new one.

A way to enhance the economic picture is to recognize that the acts of not polluting and of not using resources is worth money. However, it is difficult to place a dollar value on this.

It is important to remember that eventually even the rechargeable batteries will need to be replaced and the spent batteries should not be disposed of in the household trash.

14. *Environmental Accounting and Liability Solution 14.* [wrl]. For solid waste transportation, the following equation can be used to predict the potential liability of waste hauling:

$$PL = \frac{\$10,000}{\text{ton spilled}} \left(\frac{x \text{ tons spilled}}{\text{overturn}} \right) \left(\frac{u \text{ tons waste}}{\text{yr}} \right) \left(\frac{1 \text{ trip}}{z \text{ tons}} \right) \left(\frac{1 \text{ spill}}{4,000 \text{ trips}} \right)$$

where PL = potential liability; x = estimated tons spilled during an accident; u = total tons of waste disposed of/yr; and z = typical amount of waste loaded/truck at the plant.

For this problem, the following substitutions can be made:

$$PL = \frac{(\$10,000/\text{ton}) \ (2 \text{ tons/accident}) \ (32,000 \text{ tons disposed/yr})}{(4,000 \text{ trips/accident}) \ (4 \text{ tons/trip})}$$

PL = $40,000/yr, or on a $/ton basis:

$$\frac{\$40,000/\text{yr}}{32,000 \text{ tons/yr}} = \$1.25/\text{ton}$$

14. *Environmental Accounting and Liability Solution 14.* [wrl] (continued)

As this problem demonstrates, pollution prevention activities which reduce the generation of waste can substantially reduce the cost of future liabilities, as well as required disposal costs.

15. *Environmental Accounting and Liability Solution 15.* [cjk]. The straight–line rate of depreciation is a constant equal to 1/r, where r is the life of the facility for tax purposes. Thus, if the life of the plant is 10 years, the straight–line rate of depreciation is 0.1. This rate, applied over each of the 10 years, will result in a depreciation reserve equal to the initial investment.

16. *Environmental Accounting and Liability Solution 16.* [cjk, dj]. The payout time is calculated as the fixed capital investment divided by the sum of the annual profit plus the annual depreciation.

$$\text{Payout Time} = \frac{\text{Fixed Capital Investment}}{(\text{Annual Profit} + \text{Annual Depreciation})}$$

When the formula is applied to the data presented in the Problem Statement, the following result is obtained:

$$\text{Payout Time} = \frac{\$10,000}{(\$1,500 + \$1,430)} = 3.44 \text{ yr}$$

17. *Environmental Accounting and Liability Solution 17.* [cjk, dj]. When the percent rate of return on investment formula is applied to the data presented in Problem Statement 17, the following result is obtained:

$$\text{Percent rate of return} = \frac{\$1,500/\text{yr}}{\$10,000} \times 100 = 15\%/\text{yr}$$

This rate of return is higher than could be earned with many financial investments. The plant manager would be well advised to make the investment.

18. *Environmental Accounting and Liability Solution 18.* [cjk]. In order to select the most economical scrubber, a comparison can be done among the three scrubbers based on the Return on Investment (ROI) as defined in the Problem Statement. The following table can be used to simplify these calculations.

18. *Environmental Accounting and Liability Solution 18.* [cjk] (continued)

Table 32
Summary Cost Data for Scrubber Options A, B, and C

Scrubber	A	B	C
Capital Investment	$300,000	$400,000	$450,000
Depreciation	$30,000	$40,000	$45,000
Operating Costs	$50,000	$35,000	$25,000
Total Annual Costs	$80,000	$75,000	$70,000

A comparison between Scrubbers A and B based on the ROI calculation is carried out as follows:

$$ROI = \frac{(\$30,000/yr- \$40,000/yr)+ (\$50,000/yr - \$35,000yr)}{(\$400,000- \$300,000)}(100)$$

$$ROI = \frac{(-\$10,000/yr)+ (\$15,000/yr)}{(\$100,000)}(100) = \frac{(\$5,000/yr)}{\$100,000}(100) = 5\%$$

A comparison between Scrubbers A and C based on the ROI calculation is carried out as follows:

$$ROI = \frac{(\$30,000/yr- \$45,000/yr)+ (\$50,000/yr - \$25,000yr)}{(\$450,000- \$300,000)}(100)$$

$$ROI = \frac{(-\$15,000/yr)+ (\$25,000/yr)}{(\$150,000)}(100) = \frac{(\$10,000/yr)}{\$150,000}(100) = 6.66\%$$

Scrubber C has the highest ROI and should be selected as the most economical scrubber of the three being evaluated.

19. *Environmental Accounting and Liability Solution 19.* [fwk]. It is suggested that the instructor list the 30 locations in the first column of a spreadsheet. The second column should be used for the final sample concentrations. In the third column, enter a 90 at the hot spot, and a 30 just above and below it. In the fourth column, enter a formula which adds the value in the hot spot column to a constant background concentration of about 5 (from another cell in the spreadsheet). In the fifth column, generate random numbers for noise, in the range of zero to about 4. For example, in Lotus and Quattro, the function @RAND generates random numbers between 0 and 0.999, which would be multiplied by a constant from another cell, around 4. Now go back to your second column, and enter the sum of Columns Four and Five. Then print the results on paper.

19. *Environmental Accounting and Liability Solution 19.* [fwk] (continued)

(Remember that the numbers change every time a recalculation is carried out.) To prevent different teams from sharing their results, it might be appropriate to move the hot spot around and create a different results sheet for each team.

Each team should submit their results on a standard form. The instructor would then use the results sheet to find the sampling result for each location requested by the team, and write those results on the standard form. After receiving the results, the team might submit another round of sampling results. Eventually, the team will submit their remediation plan. The instructor then charges the team the appropriate amounts for each sample bottle, sample shipping cost, and for each laboratory analysis, plus the drilling crew charge for each round of samples, the remediation charge for the portion remediated, and the penalty charge for any portion missed. The team having the lowest overall cost would then receive the highest grade.

A team would be wise to practice a bit before submitting their actual results. One team member could create a sample problem, and the others could try to find the best solution. This would give the team a feel for the balance between over-remediating versus risking too high a penalty, and for spending too much on a large number of samples versus having to pay for another day for the drilling crew.

In deciding where to sample, careful attention might be focused on the width of the expected hot spot, remembering, however, that a slight rise in concentration might be due to random variability rather than due to the hot spot.

An extremely able team, or a team that has already been successful at this exercise, might try "composite" samples. For example, if "grabs" were taken at Locations 1 through 5, mixed together, and sent as a single sample to the lab, it would save on analysis costs. If the hot spot were hit, the results would be higher than background by one-fifth (if there are five grabs in the sample bottle) of the difference between background concentration and the hot spot concentration. This is why the neighboring farm results are given. Compositing will not be useful if the number of grabs is so high that an elevated result would be within a couple of standard deviations of the variance of clean samples. Because of the need to check the statistics when compositing, this portion of the problem would be useful when teaching statistics, or when it is important to emphasize the practical usefulness of statistical skills.

As an example of results, a team sampled all even distances, 2, 4, 6, etc. to 30. Fifteen samples cost:

$$15 \ (\$200) = \$3,000$$

19. *Environmental Accounting and Liability Solution 19.* [fwk] (continued)

plus one drilling crew day of $3,000. The remediation plan went from position 3,4 to 6,4 but missed some contamination that ran to 6,6. The remediation cost was:

$$(6,4 - 3,4) (\$4,000) = \$12,000$$

and the penalty was:

$$(6,6 - 6,4) (\$6,000) = \$1,200$$

The total cost was $19,200, which is a fairly good result.

For further information, see **Gilbert, R.O.**, Probability that a hot spot exists when none was found, in *Statistical Methods for Environmental Pollution Monitoring*, Van Nostrand, New York, 1987, 128.

Chapter 14

OTHER ENVIRONMENTAL ISSUES

Mike Haradopolis

I. INTRODUCTION

Since the dawn of time there has been a constant conflict between humans and their environment. Today society often faces the dilemma of "jobs or the environment," and society must now address the "politics" of environmental issues. In recent history the U.S. government has created numerous federal agencies designed to oversee the environment. These agencies, the lead of which is the U.S. EPA, coupled with an increased awareness of environmental dangers, have helped create cleaner lakes, streams, forests, and oceans surrounding the U.S. However, the nation still faces many potential hazards: acid rain, nuclear waste disposal, global warming, pollution from leaking underground storage tanks, and air pollution to name just a few.

Major environmental problems exist, but many fear strict environmental regulations. Leaders in business and industry warn that over-regulation may stifle economic growth necessary to sustain the present living standards of people in the U.S. Led by the U.S. EPA, federal and state governments drive the expenditure of billions of dollars each year by regulating the interaction between business and the environment. Some analysts estimate the cost of strict regulatory compliance to reach upwards of $100 billion dollars per year, expenses which are passed on to consumers and taxpayers. Thus, not all answers to environmental concerns are clear-cut.

This chapter examines environmental issues from a business perspective as well as from the environmentalist approach. The net effect of a strict regulatory climate on U.S. business is not simple to determine as some businesses may have large expenditures associated with environmental compliance, while many others that assist in environmental compliance and remediation experience business growth as a direct result of strict regulation.

There is no substitute for a clean and healthy environment. However, many businessmen and civil rights activists fear that too much government regulation will jeopardize personal liberties and basic, constitutional private property rights.

This chapter presents to the reader these basic issues and information on some not-so-obvious environmental problems. Through an analysis of problems contained in this chapter the reader will obtain a clear understanding of the basic issues and

303

past problems to better grasp the nature of pressing environmental concerns and environmental dilemmas facing the U.S. in the not too distant future.

II. PROBLEMS

1. *Other Environmental Issues Problem 1.* (capital costs, operating costs, environmental costs). [mh]. Describe and discuss the basic changes made in project cost estimating practices of companies in the U.S. from 1935 to today due to environmental regulations and resource management concerns.

2. *Other Environmental Issues Problem 2.* (pollution prevention, sustainable development, economic development). [sl]. What is the difference between sustainable economic development and pollution prevention?

3. *Other Environmental Issues Problem 3.* (Clinton, U.S. EPA, Gore). [mh]. The Clinton-Gore administration believes the U.S. has made tremendous strides in improving the environment in the last 25 years. Discuss some of the improvements made in environmental conditions in the U.S. over the past 25 years.

4. *Other Environmental Issues Problem 4.* (CFC, Title VI, Montreal Protocol). [pcy]. Briefly describe the Montreal Protocol regarding CFCs.

5. *Other Environmental Issues Problem 5.* (U.S. EPA, budget, controversy). [mh]. In the 1980s the U.S. EPA went through a series of major problems. These problems included budget cuts and alleged misappropriations of funds. Discuss these budget cuts and the controversial figures involved in EPA at the time.

6. *Other Environmental Issues Problem 6.* (industrial solid waste management, hazardous waste management, Superfund, Love Canal). [sn]. The Love Canal is a hazardous dump site located in the industrial community of Niagara Falls, New York. It is the most notorious waste site in U.S. history, having led to the passage of the Superfund legislation in 1980. It symbolizes a tragedy that finally compelled assertive government action to correct decades of waste disposal abuses. The incident has had a significant impact on current hazardous waste practices and management.
Since the Love Canal incident, what changes have been observed in the attitude of government, industry and the general public toward management of industrial solid and hazardous waste?

7. *Other Environmental Issues Problem 7.* (U.S. EPA, asbestos, workplace standards). [rt]. Over the last 15 years, the U.S. EPA and several other federal agencies have acted to prevent unnecessary exposure to asbestos by prohibiting some of its uses and by setting exposure standards in the workplace. Five agencies

7. *Other Environmental Issues Problem 7.* [rt] (continued)

have major authority to regulate asbestos. Name these five agencies and briefly explain how each regulates asbestos.

8. *Other Environmental Issues Problem 8.* (biosphere, biogeochemical cycles). [pcy]. What is the biosphere and what are biogeochemical cycles?

9. *Other Environmental Issues Problem 9.* (acid rain, SO_2, NO_2). [pcy]. How does acid rain form and what are the causes and effects of acid rain?

10. *Other Environmental Issues Problem 10* (greenhouse effect, CFCs). [pcy]. What is the greenhouse effect? How does ozone depletion occur in the stratosphere?

11. *Other Environmental Issues Problem 11.* (EMF, leukemia). [pcy]. What are EMFs? Are there any standards setting maximum EMF exposure levels and what are the potential health effects of EMF exposure? What have studies of potential health effects from electromagnetic fields (EMFs) shown since 1980?

12. *Other Environmental Issues Problem 12.* (UST, RCRA). [pcy]. Define the term underground storage tank (UST). List the types of tanks which are not regulated by the Resources Conservation and Recovery Act (RCRA) UST program. Also list the major causes of tank failures.

13. *Other Environmental Issues Problem 13.* (USTs, RCRA, U.S. EPA). [mh]. Discuss in general terms environmental issues related to underground storage tanks (USTs).

14. *Other Environmental Issues Problem 14.* (nuclear waste, radioactive waste, HLW, TRU, LLW). [pcy]. Briefly explain the following radioactive waste terms:

    a. Defense Wastes
    b. Commercial Wastes
    c. High Level Wastes (HLW)
    d. Transuranic Wastes (TRU)
    e. Low Level Wastes (LLW)

15. *Other Environmental Issues Problem 15.* (infectious waste, hospital waste, medical waste). [pcy]. What are major sources of infectious wastes? Define regulated medical waste. List the treatment methods for hospital waste.

16. *Other Environmental Issues Problem 16.* (noise pollution, hearing protection). [rt]. It is estimated that between 8.7 and 11.1 million Americans suffer from permanent hearing disability, yet it is almost impossible for an active person to avoid exposure to potentially harmful sound levels in today's mechanized world. Explain the effect of environmental noise pollution. What may individuals do to reduce noise effects in their everyday lives?

17. *Other Environmental Issues Problem 17.* (transuranic, WIPP, DOE). [rt]. A transuranic (TRU) waste program site has been located at the Waste Isolation Pilot Plant (WIPP) in Carlsbad, New Mexico. The WIPP facility is a Department of Energy (DoE) research and development facility that has been designed to accept 6 million ft³ of contact-handled TRU waste, as well as 25,000 ft³ of remote-handled TRU waste. The facility will accept defense-generated waste and place it into a retrievable geologic repository. A geological repository is, in this instance, the salt formations located near Carlsbad. The facility has a design-based lifetime of 25 years. However, a recent earthquake (5.3 on the Richter scale) shook the area near this facility.

From an engineering standpoint, do you feel this is a safe TRU site? Defend your answer.

Reference: You may find help with your answer on the Internet at www.wipp.carlsbad.nm.us.

III. SOLUTIONS

1. *Other Environmental Issues Solution 1*. [mh]. Although technical parameters influencing the selection and design of a given engineering system may be unique, cost is the only parameter common to all systems. Cost is generally the main parameter used to select the optimum system from the alternatives available. Environmental regulations have significantly changed standard cost accounting practices over the years as costs of environmental impacts, energy use, and health and safety compliance have grown.

Before 1945, capital costs were the only things generally considered in the design and selection of a particular system. There are many things that go into capital costs. These may include: real estate, equipment, installation, taxes, freight, design, and start-up costs. Design and start-up costs are usually a one-time cost, with the remainder being on-going costs.

After 1945, operating costs began to be included in cost analyses for process design and selection. These costs included raw materials, operation and maintenance, utilities, energy, labor, depreciation, shipping, real estate, and taxes. These costs added yearly operation costs to the capital costs of projects, affecting process selection between low capital/high O&M cost projects versus high capital/low O&M cost projects.

By 1970, environmental costs began to grow in importance through government regulation and public pressure. These costs may include: water, wastewater, and air pollution control costs; solid waste management costs; remediation costs; and health and safety considerations.

Finally in 1990, hidden costs were uncovered and companies were then liable for retroactive litigation (ex post facto). These costs may include: regulatory costs, liability, limited borrowing power, stockholder support, consumer relations, and worker support.

With compliance and liability costs now increasingly being incorporated into cost estimates it is becoming more possible to estimate the true life-cycle cost for all processes so that a true, long-term, least cost option can be selected.

2. *Other Environmental Issues Solution 2*. [sl]. Sustainable economic development represents the set of actions that meet the needs of the present without compromising the ability of future generations in meeting their own needs.

Pollution prevention represents the set of actions that involve the reduction, to the greatest extent feasible, of generated waste. Pollution prevention has two main objectives:

- The reduction of the total quantity or volume of waste.
- The reduction of toxicity of the waste.

2. *Other Environmental Issues Solution 2*. [sl] (continued)

These reductions are carried out to minimize the present and future threats of the waste to human health and the environment.

3. *Other Environmental Issues Solution 3*. [mh]. Since the first Earth Day 25 years ago, the American people have seen improvements in public health, worker safety, and the natural environment. The U.S. has taken the lead out of gasoline and paint. The U.S. has virtually eliminated direct discharge of raw sewage into the nation's water supply. The nation has banned DDT and other dangerous and persistent pesticides. Because of these and other actions, lead levels in the average American's bloodstream have dropped by 25% since 1976. In addition, U.S. citizens can now fish and swim in formerly polluted waters, and the bald eagle has been removed from the endangered species list.

4. *Other Environmental Issues Solution 4*. [pcy]. In 1987, the U.S. and 22 other nations signed the Montreal Protocol. These nations agreed to limit the production and use of chloroflurocarbons (CFCs), which are believed to deplete the upper ozone layer. The initial protocol required 50% reduction in CFC production worldwide by the year 2000. Now signed by more than 140 nations, the Protocol was amended to require a virtual phaseout of CFC production by January, 1996.

The Montreal Protocol is also the document that formed the basis for Title VI of the U.S. Clean Air Act of 1990.

5. *Other Environmental Issues Solution 5*. [mh]. President Ronald Reagan appointed Anne Buford to head the U.S. EPA in 1981. She sought to ease enforcement of what Reagan's administration perceived as overly stringent environmental regulations. She also sought to emphasize more voluntary compliance from business and proposed to reduce budget expenditures for the EPA. In fact, in the first 2 years of the Reagan administration the EPA reduced its proposed budget by over 45%.

During this period, the agency experienced many problems. Rita Lavalle was selected to head CERCLA (Superfund). She was accused of funneling Superfund monies to Republican Congressional districts. Ultimately, she was found guilty of obstructing a Congressional investigation and was convicted of perjury. In 1983 she was sentenced to prison. In the wake of this scandal, Anne Buford resigned and was replaced by William Ruckelshaus, a former head of the EPA in the 1970s.

6. *Other Environmental Issues Solution 6*. [sn]. The Love Canal incident occurred in the mid-1970s. Since then, hazardous waste problems have led the public to initiate action and have

6. *Other Environmental Issues Solution 6.* [sn] (continued)

brought attention to government agencies, scientists, and professionals. There are quite a number of local, state, and national organizations involved with the management of hazardous waste. A successful example of a national organization which grew out of Love Canal is the "Citizens' Clearing House for Hazardous Waste." This organization acts as a resource center to provide community residents with scientific information on a variety of topics including the toxic effects of chemicals, landfill technology, and alternate disposal of hazardous waste. Other services include linking people and relevant government agencies together, helping communities organize, and referring people to sources of technical and legal help.

In some cases, a community health profile of people living near hazardous waste disposal sites has been initiated to collect information about demographics and health problems. This has been found to be useful to public health professionals in planning risk assessment studies of hazardous waste sites. It was also found that people who live near a hazardous waste disposal site have more concerns about their health and are more educated about the toxicity of chemicals that have been found on the site than people who are far removed from these disposal areas.

From a government perspective, after the Love Canal incident, government and state agencies have come to realize that mismanagement of hazardous waste has created substantial risks to human health and the environment. The Resource Conservation and Recovery Act (RCRA) was enacted in 1976 and was later amended in 1984 (the Hazardous and Solid Waste Amendments of 1984) to deal with the problems of solid and hazardous waste. Since 1976, hazardous wastes have been regulated based on the "cradle to grave" concept. Under these laws, the government requires existing and future treatment, storage, and disposal facilities to maintain operating records and develop a groundwater monitoring program during operation and post-closure periods of disposal.

Financial responsibility requirements for hazardous waste facilities were also promulgated by the EPA to prevent the abandonment or improper closure of hazardous waste facilities and their attendant hazards. The government also came to realize the need for resources for the clean-up of abandoned sites, which later led to the passage of the Comprehensive Environmental Response, Compensation and Liability Act of 1980 (CERCLA) or "Superfund." Since then the disadvantages of traditional pollution control practices have become evident. The environmental management of hazardous waste has been shifting toward a pollution prevention approach which is designed to limit the amount of waste generated in the first place, and to ensure that the wastes pose no threat to

6. *Other Environmental Issues Solution 6.* [sn] (continued)

people's health or the environment. The Pollution Prevention Act of 1990 established pollution prevention as a national policy, and emphasized its role as the lead management tool within the hierarchy of waste management approaches.

Numerous state and municipal governments enacted worker Right-to-Know laws which led to the promulgation of a Federal Community Right-to-Know law entitled the Hazardous Communication Standard, 29 CFR 1910.1200. The objective of this Act is to communicate to the worker and to the community the presence and effects of hazardous chemicals in the workplace and within commercial facilities in their community.

From an industry point of view, many environmental laws have been formulated and enacted that led to complex administrative requirements for record-keeping, conformance to manifest systems, development of plans and training programs, securing facility permits, etc. Also, industry has to be fully aware of requirements for owners and operators of treatment, storage, and disposal facilities which include the availability of funds to cover closure costs for the facility in an environmentally sound manner as addressed in the Superfund Amendments and Reauthorization Act (SARA) of 1986. Pollution prevention clearly has become a top priority for industries since it reduces the cost of waste management, regulatory compliance, liabilities, and other indirect short-term and long-term costs of doing business.

7. *Other Environmental Issues Solution 7.* [rt]. The five agencies having major authority over the regulation of asbestos include:

1. The Occupational Safety and Health Administration (OSHA) which sets limits for worker exposure on the job.
2. The Food and Drug Administration (FDA) which is responsible for preventing asbestos contamination in food, drugs, and cosmetics.
3. The Consumer Product Safety Commission (CPSC) which regulates asbestos in consumer products.
4. The Mine Safety and Health Administration (MSHA) which regulates mining and milling of asbestos.
5. The U.S. Environmental Protection Agency (U.S. EPA) which regulates the use and disposal of toxic substances in air, water, and land, and has banned all uses of sprayed asbestos materials. In addition, EPA has issued standards for handling and disposing of asbestos-containing wastes.

8. *Other Environmental Issues Solution 8.* [pcy]. The biosphere is defined as that part of the planet that sustains life. It encompasses the lower part of the atmosphere, the hydrosphere (oceans, lakes, rivers and streams), and the lithosphere (the earth's crust) down to a depth of approximately 2 km.

Biogeochemical cycles are transport pathways, and the chemical and physical interactions of the elements within and among these regions of the biosphere.

9. *Other Environmental Issues Solution 9.* [pcy]. Acid rain with a pH below 5.6 is formed when certain anthropogenic air pollutants travel into the atmosphere and react with moisture and sunlight to produce acidic compounds. Sulfur and nitrogen compounds released into the atmosphere from different sources are believed to play the biggest role in formation of acid rain. The natural processes which contribute to acid rain include lightning, ocean spray, decaying plant and bacterial activity in the soil, and volcanic eruptions. Anthropogenic sources include those utilities, industries, businesses, and homes that burn fossils fuels, plus motor vehicle emissions. Sulfuric acid is the type of acid most commonly formed in areas that burn coal for electricity, while nitric acid is more common in areas that have a high density of automobiles and other internal combustion engines.

There are several ways that acid rain affects the environment:

- Contact with plants can harm plants by damaging outer leaf surfaces and by changing the root environment.
- Contact with soil and water resources. Due to the acid in the rain, fish kills in ponds, lakes and oceans, as well as effects on aquatic organisms, are common. Acid rain can cause minerals in the soil to dissolve and be leached away. Many of these minerals are nutrients for both plants and animals.
- Acid rain mobilizes trace metals, such as lead and mercury. When significant levels of these metals dissolve from surface soils they may accumulate elsewhere, leading to poisoning.
- Acid rain may damage building structures and automobiles due to accelerated corrosion rates.

The general chemical formulae for the formation of acid rain are as follows:

$$SO_x + O_2 \rightarrow SO_2 + H_2O \rightarrow H_2SO_4$$
$$NO_x + O_2 \rightarrow NO_2 + HNO_3$$
$$CO_2 + H_2O \rightarrow H_2CO_3$$

10. *Other Environmental Issues Solution 10.* [pcy]. Fossil-fuel burning, forestry and agricultural practices are responsible for most of the anthropogenic contributions to the gases in the atmosphere. These gases, primarily carbon dioxide and methane, are transparent to short-wave radiation, but absorb long wavelength light energy as it reflects off the earth's surface back into space. This selective absorption of light results in an increase in temperature of the lower atmosphere high in these "greenhouse gases," with the transfer of this absorbed energy to the earth's surface, hence the greenhouse effect. This is a major cause of the global warming problem. Carbon dioxide, emitted to air during combustion, is the major chemical contributing to the problem.

In 1974, F. Sherwood Rowland and Mario Malina, both in the chemistry department of the University of California at Irvine, released a study proposing that CFCs diffused through the troposphere into the stratosphere, altering the chemistry of the protective ozone layer. The CFC gases were first developed by chemists at General Motors Corporation in the early 1930s. They were thought to be almost perfect chemicals because they are stable and do not react with other substances. CFCs were used in aerosol sprays for hundreds of different kinds of consumer products. CFCs also replaced toxic and flammable gases once used as coolants in refrigerators. Since CFCs are non-flammable, non-toxic and non-corrosive, they can be used in a variety of products without undergoing drastic changes in their properties and without the threat of fire or other hazards. However, the stability of CFCs allow them to survive for many years and accumulate in the troposphere. Some CFCs eventually move into the stratosphere, where they can deplete the ozone layer there.

11. *Other Environmental Issues Solution 11.* [pcy]. Electromagnetic fields (EMFs) are invisible lines of force surrounding any electrical device. Power lines, electrical wiring and appliances all produce EMFs. Electric fields are produced by voltage, measured in volts per meter (v/m) or kilovolt per meter (kv/m), and are easily shielded by conducting objects like trees and buildings. Electric fields decrease in strength with increasing distance from the source. Magnetic fields are produced by current, measured in gauss (G) or tesla (T), and are not easily shielded by most materials. Magnetic fields decrease in strength, as electric fields do, with increasing distance from the source.

There are no federal health standards and regulations related to EMF exposure. Six states have set standards for transmission line electric fields (Florida, Minnesota, Montana, New Jersey, New York and Oregon). The states of New York and Florida have also set standards for transmission line magnetic fields. To date, 14 studies have analyzed a possible association between proximity to power lines and various types of childhood cancer. Four of the 14

11. *Other Environmental Issues Solution 11.* [pcy] (continued)

studies showed a statistically significant association with acute lymphocytic leukemia, the most common form of leukemia.

The most frequently reported health effect of EMFs is cancer, particularly elevated risks of leukemia, lymphoma, and nervous systems cancer in children. Also, birth defects, behavioral changes, slowed reflexes, and spontaneous abortions have been noted.

12. *Other Environmental Issues Solution 12.* [pcy]. An underground storage tank (UST) is defined as any storage tank with at least 10% of its volume buried below ground, including pipes attached to the tank. Aboveground tanks with extensive piping may be regulated under the Resource Conservation and Recovery Act (RCRA) Sub-Title I, UST regulations. Types of tanks to which the UST program does not apply include:

- Farm and residential tanks which store less than 1,100 gallons of motor fuel
- On-site heating oil storage tanks
- Septic tanks and sewers
- Pipelines for gas or liquid
- Surface impoundments
- Flowthrough process tanks

Major UST tank failures are due to:

- Corrosion
- Faulty installation
- Pipe failure
- Overfills

13. *Other Environmental Issues Solution 13.* [mh]. Under RCRA USTs are defined as tanks with 10% or more of their volume, including piping, underground. The main purpose of USTs is to reduce potential damages from accidental releases of flammable liquids stored on the surface. Five to six million USTs which contain hazardous substances or petroleum are located in the U.S. It is estimated that approximately 10% of all USTs are leaking and that many more will leak in the future. Products released from these leaking tanks may threaten groundwater supplies, damage sewer lines and buried cables, poison crops, and lead to fires and explosions. Studies done in 1985 revealed that around one third of existing motor fuel storage tanks were over 20 years old and, of these, most were constructed of steel and were not protected against corrosion. The primary reason for regulating USTs is to protect the groundwater for human consumption.

13. *Other Environmental Issues Solution 13.* [mh] (continued)

The biggest challenge in the design of a UST is to achieve a better, cheaper tank design and improve leak detection technology. Federal, state and local laws require owners to register their tanks and indicate their age, location, and content. Owners have to detect leaks from new and existing tanks, clean up environmentally harmful releases from them, and litigate third party damages resulting from such leaks.

The U.S. EPA has three sets of regulations pertaining to USTs. The first addresses technical requirements for petroleum and hazardous substance tanks. The second involves legislation that addresses financial responsibility requirements for underground petroleum tanks, The third addresses standards for approval of state tank programs.

14. *Other Environmental Issues Solution 14.* [pcy].

    a. Defense Wastes: Those wastes which have been generated over the period during and since World War II, at three main Department of Energy (DoE) installations: the Hanford Site near Richland, Washington; the Idaho National Engineering and Environmental Laboratory, near Idaho Falls, Idaho; and the Savannah River Plant near Aiken, South Carolina.

    b. Commercial Wastes: Those wastes which are produced by reactors used for the generation of electric power, by facilities used to process reactor fuels, and by a variety of institutions and industries.

    c. High Level Wastes (HLW): Those wastes resulting from reprocessing of spent fuel or are the spent fuel itself, either of a DoD or commercial origin.

    d. Transuranic Wastes (TRU): Those wastes containing isotopes above uranium in the periodic table of elements. They are the by-products of fuel assembly, weapons fabrication and of reprocessing operations. Their radio-activity level is generally low, but since they contain several long-lived isotopes, they must be managed separately.

    e. Low Level Wastes (LLW): Those wastes officially defined as all wastes other than those defined above. The bulk of LLW has relatively little radioactivity and contains practically no transuranic elements. Most of the LLW

14. *Other Environmental Issues Solution 14.* [pcy] (continued)

       requires little or no shielding, may be handled
       by direct contact, and may be buried in near-
       surface facilities.

15. *Other Environmental Issues Solution 15.* [pcy]. The major sources of infectious wastes are from human and animal clinics, funeral homes, health care facilities, dental offices, laboratories, animal research facilities and research institutions.

In March 24,1989, the U.S. EPA published regulations in the Federal Register as required under the Medical Tracking Act of 1988. The term "medical waste" was defined as

       "...any solid waste that is generated in the diagnosis,
       treatment, or immunization of human beings or animals, in
       research pertaining there to, or in the production or testing
       of biologicals."

Seven classes of listed medical wastes are defined by the U.S. EPA as follows:

- Cultures and stocks
- Pathological waste
- Human blood and blood products
- Sharps (needles)
- Animal wastes
- Isolation wastes
- Unused sharps

There are six common treatment techniques used for hospital wastes. These treatment techniques include:

- Incineration
- Autoclaving
- Microwaving
- Chemical disinfection process
- Irradiation
- Plasma system

16. *Other Environmental Issues Solution 16.* [pg]. Noise causes temporary or permanent hearing loss, physical and mental disturbances, breakdowns in the reception of oral communications, reduced efficiency in performing work-related tasks, irritability, disruption of sleep and rest, and increased potential for accidents. Steady exposure to 90 dB of sound can cause eventual hearing loss. Heavy traffic can reach 90 dB, and the noise of jet planes, rock-and-roll bands, motorcycles, power mowers, auto horns, heavy construction, and farm equipment exceed this level. Noise must be

16. *Other Environmental Issues Solution 16.* [pg] (continued)

suppressed, since it is a source of irritation, distraction, and emotional strain, and fosters inefficient performance.

The following is a list of noise protection measures that an individual may employ to prevent long-term hearing loss:

- Wear ear muffs and/or ear inserts.
- Be aware of major noise sources near any residence.
- Look for construction methods/techniques that produce a quieter local environment (wall-to-wall carpeting, wall and door construction, insulating the heating and air-conditioning ducts, etc.).
- Compare the noise outputs of different makes of appliance models before making a selection.

17. *Other Environmental Issues Solution 17.* [rt]. Even though an earthquake did occur, this is still a safe TRU site for the following reasons:

- Salt deposits demonstrate the absence of flowing fresh water that could move waste to the surface. Water, if it had been or were present, would have dissolved the salt beds.
- Rock salt heals its own fractures because of its plastic quality. That is, salt formations will slowly and progressively fill in mined areas and safely seal radioactive waste from the environment.
- Salt rock also provides shielding from radioactivity similar to that of concrete.
- The primary salt formation containing the WIPP mine is about 2,000 feet thick, beginning 850 feet below the surface.

Index

A

acceptable risk 44, 49, 52, 62

accident prevention 197-198, 204

acid rain 148, 157, 306, 312

acronyms 7, 19-21, 287-288, 294

acute effects 28, 36

adsorption 153, 168

advanced waste treatment 148, 159

air exchange rates 116, 126-130

air pollution 6-7, 19
    law 7, 19
    regulations 7, 19

air quality issues 113-144
    models 31-32, 40-41
    standards 6, 19

ambient air quality 116, 126

ammonia 155-156, 172-176

animal waste management 155-156, 172-176

ANSI 218, 221

anthracene 120-121, 137-138

anthropometry 198, 207

ARARs 4, 12

asbestos 117, 131-132, 305-306, 311

atrazine 29, 38

attitudes 259-260, 269-270

audits 219, 226-227, 258-260, 268-271

automobile 96, 107-108

average person 28, 36

averaging time 28, 36

B

BACT 8, 21-22

bag houses 123, 140-141

banks 287, 294

base closure 240, 252-253

batteries 290, 297-298

benzene 29-30, 39, 120-121, 137-138, 155, 171-172

BIA 240-241, 253

biodegradation 155, 171-172

biogeochemical cycles 306, 312

biological hazards 198, 205

biomechanics 198, 207

biosphere 306, 312

biochemical oxygen demand (BOD) 148-149, 151-152, 159-160, 166

Black Hills 238-239, 251

BLM 240-241, 253

break-even costs 70, 79

budget 6, 18, 305, 309